T0357580

RANSOM WAR

MAX SMEETS

Ransom War

How Cyber Crime Became a Threat to National Security

OXFORD
UNIVERSITY PRESS

OXFORD
UNIVERSITY PRESS

Oxford University Press is a department of the
University of Oxford. It furthers the University's objective
of excellence in research, scholarship, and education
by publishing worldwide.

Oxford   New York

Auckland   Cape Town   Dar es Salaam   Hong Kong   Karachi
Kuala Lumpur   Madrid   Melbourne   Mexico City   Nairobi
New Delhi   Shanghai   Taipei   Toronto

With offices in

Argentina   Austria   Brazil   Chile   Czech Republic   France   Greece
Guatemala   Hungary   Italy   Japan   Poland   Portugal   Singapore
South Korea   Switzerland   Thailand   Turkey   Ukraine   Vietnam

Oxford is a registered trade mark of Oxford University Press
in the UK and certain other countries.

Published in the United States of America by
Oxford University Press
198 Madison Avenue, New York, NY 10016

Library of Congress Cataloging-in-Publication Data is available
Max Smeets.
Ransom War: How Cyber Crime Became a Threat to National Security.
ISBN: 9780197803035

Printed in the United Kingdom on acid-free paper
by Bell and Bain Ltd, Glasgow

For Madelyn

CONTENTS

INTRODUCTION

When employees of the Costa Rican Ministry of Finance returned to work after a prolonged Easter weekend on April 18, 2022, they were greeted by an unpleasant surprise. Their organization had become victim of a ransomware attack, a type of malicious activity where hackers lock access to files or systems until a ransom is paid.[1] The initial message posted on April 17 from the perpetrators was short: "We downloaded 1 TB of your portal databases as well as internal documents, we will start publishing this data on April 23".

A criminal group calling themselves 'Conti' boldly announced their involvement through their leak site, a public website used to publish and advertise the data stolen from their victims. In an update to their initial post, Conti placed an initial ransom demand of $10 million, a demand that was subsequently rejected by the Costa Rican government, stating that they were not willing to pay any money to an organized criminal group.[2] Conti accompanied their demand with the claim that their original post on the leak site had actually come before they encrypted the data of the Ministry of Finance. In an embarrassment to the government, this suggested that the encryption could have been thwarted if it had been detected sooner.[3] In fact, it was later revealed that Conti had been covertly present in the Ministry of Finance's networks for nearly a week before the breach was detected. The cybercriminal group initially breached the system on April 11 via a compromised VPN access—a relatively straightforward method[4]—and had exfiltrated up to 672 Gigabytes of data by April 15, before it encrypted the

1

ministry's data as Ministry employees logged off to enjoy the long Easter weekend.[5]

Conti's public post signaled the beginning of an extended ransomware campaign targeting various Costa Rican government institutions. Around the same time as the intrusion into the Ministry of Finance's systems was discovered, a website belonging to the Ministry of Science, Innovation, Technology and Telecommunications was compromised and defaced.[6] Then, the group gained access to the Costa Rican Social Security Fund's (CCSS) account on X, formerly Twitter, and posted several messages before the password was changed and the CCSS regained control in the early morning hours of Tuesday, April 19.[7] On the same day, it was discovered that Conti had also hacked into and stolen information from the email servers of the National Meteorological Institute as well as the state-owned internet provider Radiográfica Costarricense, along with subsequent hacks affecting Alajuela Interuniversity, the Ministry of Labor and Social Security, the Social Development and Family Allowances Fund, and attempts at the Rural Development Institute.[8] Additionally, JASEC, a local electricity provider in Cartago, saw its administrative systems compromised and encrypted.[9] A Costa Rican official later stated that in total at least twenty-seven institutions had been affected by Conti's assault.[10]

Conti's repeated cyberattacks ultimately led to the President of Costa Rica issuing an executive order on May 8, 2022. This order declared a national emergency in response to the ransomware attacks on its public sector and pronounced that the country was in a "state of war".[11] This marked the first time in history that a country formally declared a state of emergency in direct response to ransomware, or in fact any cyberattack. The country was now effectively engaged in a 'ransom war'.[12]

Early on, a variety of international partnerships, spanning both governmental and private sectors, committed to helping Costa Rica in addressing the cyber threat it encountered. This group included countries such as Spain, the United States and Israel, alongside corporate entities like Microsoft and GBM, a prominent IT service provider in Central America.[13] Already on May 6, the

United States (US) State Department announced a $15 million bounty for information leading to the identification and apprehension of Conti members.[14] Conti frequently expressed their disdain for the U.S. government, accusing them of hypocrisy for instructing the Costa Rican Government not to pay the ransom while having purportedly paid ransoms themselves in the past.[15] Moreover, Conti disparaged the United States as "a cancer on the body of the earth" and mockingly called President Biden an "old fool [who] will soon die".[16]

However, before these international partnerships could get to work, the Costa Rican government itself had to do some initial damage control. When Conti's intrusion was first discovered at the Ministry of Finance in the early morning hours of April 18, it was decided that several of its systems, including the digital tax system (ATV), the salary payment platform (Integra II), as well as the digital customs control system (TIC@), would be shut down to make it easier to conduct investigations.[17] The tax system being down that day (and many days to follow) was especially problematic, as different kinds of taxes such as the VAT were due on Monday, April 18.[18] Not being able to access the state's tax system made it hard for citizens and companies to fulfill their duties. Consequently, the minister of finance, Elián Villegas, postponed the due date until the systems would be fully restored. It would take almost two full months until that was the case on June 13.[19]

Taking down the Integra II system also brought about considerable challenges. The Ministry of Public Education (MEP), which uses the Integra II system, has the largest payroll in the country with about 86,000 employees, most of whom are teachers.[20] In light of the loss of all access to current salary information, the Ministry of Finance's contingency plan provided to the Ministry of Public Education envisaged a manual handling of salaries, basing upcoming payments on what had been paid the months before— not including any recent pay raises or other salary adjustments. Unsurprisingly, the manual handling of this enormous payroll led to several faulty or missing payments, which affected around 12,000 teachers.[21] Naturally, this stirred up discontent among the teachers, who, even before Conti's attack, had been complaining about the Ministry's flawed handling of their salaries. Thus, these additional missing salaries led some of them to gather in protest in front of the MEP headquarters on May 17.[22] The Integra II system was finally reinstated a month later.

The effects of shutting down the TIC@ system also had severe consequences, especially at borders and airports. There, the unavailability of customs information paralyzed the cross-border movement of goods immediately.[23] Although the Ministry of Finance quickly announced a contingency plan for the processing of imports and exports, the alternative process, which included obtaining customs declarations via email, was not satisfactory.[24] In fact, the situation was so concerning that two weeks after the incident, on May 3, seven chambers of commerce jointly appealed to the government that the significantly delayed import procedure was threatening a collapse of the Costa Rican manufacturing sector.[25] It would take another month until TIC@ was finally up and running again on June 24, sixty-six days after it was shut down.

The challenges faced by Costa Ricans in the wake of the cyberattacks extended beyond just the disruptions to public sector services. They also faced a secondary effect: other cybercriminals seemed to jump on the opportunity and started to send fake requests for password-resets in the name of the Ministry of Finance. In response, the institution issued a clarification to the public, advising that these requests were not legitimate and should be ignored.[26]

While Costa Rican officials tried to contain and combat the crisis, Conti continued to taunt the government and comment on their activities via the leak site. The regular post updates revealed that the group kept a close eye on the news coverage about its activities.[27] They specifically took issue with one report by the Costa Rican newspaper *La Nación*, which drew parallels between the attacks on Costa Rica and the Distributed Denial of Service (DDoS) attacks against Estonia in the spring of 2007.[28] Apparently, Conti did not like the implication that they are affiliated with the Russian state, and they kept insisting that their sole goal is to make money and that they had no connections to any government.[29] Then again, in one of the later posts they suddenly started claiming that they aspired to overthrow the Costa Rican government and were calling on the people to start rallies and pressure their government to pay the ransom to "stabilize the situation".[30] This post wording was changed again within less than a day.[31] The last update came on May 20, simply stating that they would no longer wait for a ransom and delete the decryption key on the following Monday, May 23.[32] After that, Conti went silent while Costa Rica was left struggling with the effects of criminal group's actions. It would take several months to recover from it and deal with the loss of data.[33]

A Threat to Human and National Security

The attack on the government of Costa Rica serves as a stark example of how ransomware has become a threat to national security. In 2022, the majority of the U.K.'s government's crisis management "Cobra" meetings were convened in response to ransomware incidents rather than other national security emergencies.[34] The National Cyber Security Center in the U.K. wrote in their annual review for 2022 that eighteen ransomware incidents required a response coordinated at the national level.[35] This included a ransomware attack targeting the computer systems of water supply company South Staffordshire Water in the middle of a drought.[36] The European Union Cybersecurity Agency, ENISA, lists ransomware as one of the top fifteen threats to European citizens.[37]

Unfortunately, it is often the most vulnerable who are victim of these attacks. One chilling example took place in March 2023, when Lehigh Valley Health Network, a healthcare network in Pennsylvania, was targeted by a criminal ransomware group. When the healthcare organization refused to pay to have their data decrypted, the hackers resorted to a despicable act, leaking personal data and photos of topless female breast cancer patients.[38] In February 2024, Change Healthcare, a division of UnitedHealth Group, experienced a significant ransomware attack leading to the disruption of the largest healthcare payment system in the United States.[39] Subsequently, the health care provider confirmed that it paid $22 million in response to the ransomware demand, but patient data was still leaked.[40] In the same month, a ransomware attack impacted more than a hundred healthcare facilities in Romania, compelling some doctors to revert to using pen and paper.[41] A study based on Medicare administrative claims data revealed that during the initial week of a ransomware attack, hospital volumes drop by 17–25 percent, and that these attacks are associated with higher in-hospital mortality rates for patients admitted at the time of the attack.[42]

Policy experts and academics have written extensively on how the intelligence and military services of various governments establish offensive cyber programs to help advance their national interests.[43] It has led to lengthy debates as to whether these countries can use cyber operations to coerce states or obtain other strategic advantages.[44] Militaries also conduct numerous cyber exercises and wargames simulating the potential hacking activities of state adversaries. For example, during the world's largest and most complex international cyber defence exercise, called Locked Shields, teams from NATO, member countries and partners compete against each other to try to defend against a fictional country's attempt to execute a major cyber campaign.[45]

With the rise of ransomware, profit-driven hackers deserve at least equal consideration when it comes to safeguarding global society from cyber threats. At major hacker conference Black Hat, Chris Krebs—who served as the first Director of the Cybersecurity and Infrastructure Security Agency in the US Department of Homeland Security (DHS)—said we have "kind of over-*fetishized the advanced persistent threat*" (APT), the most advanced hacking groups, and the focus on nation-state activity.[46] According to Sami Khoury, the head of the Canadian Center for Cyber Security, the threat from nation-states remains significant, but cybercrime, of which ransomware is the most disruptive form, is "the number one cyber threat activity affecting Canadians".[47]

Yet still far too little is known about the workings of cybercriminal groups deploying ransomware and what governments and organizations can do to address this threat. The media often sensationalizes the activities of ransomware groups, highlighting their audacious criminal actions and the extent of disruption caused. The tragic impact of such events becomes the primary focus. Books authored by experts in the private sector delve into the technical aspects of operations and tend to concentrate on how defenders of specific organizations, like the Chief Information Security Officer (CISO) of a Fortune 500 company, can counter ransomware activities.

What remains largely absent in public discussions is a deeper comprehension of the organizational dynamics behind ransomware.[48] This is a critical gap that needs to be addressed. Ransomware has evolved into a form of organized crime, unlike its previous incarnation just a few years ago. These criminal groups now function akin to legitimate businesses, employing increasingly specialized processes.[49]

Moreover, an ill-understood but fundamental aspect of ransomware groups is their need to overcome what I term the *Ransomware Trust Paradox*. Despite their inherently deceiving activities—breaking into systems, stealing data, and encrypting vital information—ransomware groups must convince their victims of their trustworthiness. This trust encompasses not just the promise not to release the stolen data but also the assurance that payment will result in the decryption of the affected systems. This makes branding and reputation-building activities not peripheral but *central* to ransomware's operational success. Nation-state hacking groups—and especially intelligence agencies—aim for secrecy and ambiguity, avoiding detection and typically denying attribution if uncovered. Conversely, ransomware groups start covertly but later embrace self-attribution to enhance their brand equity. Unlike nation-state hacking groups, who shun publicity, ransomware groups often benefit from media exposure, using it to strengthen their reputation within the cybercriminal ecosystem and to the wider public. This understanding of ransomware necessitates a fundamental shift in how we approach its countermeasures.

Objectives

These crucial questions demand investigation: What are the underlying principles that drive the profitability and effectiveness of ransomware, elevating it to a significant threat to national security?

How are criminal ransomware groups organized and coordinated? And how can we effectively halt their progression?

This book collects the findings of my search to answer these questions. I have several objectives in writing this book. The first aim of the book is to systematize our knowledge about ransomware. Existing writing on specific ransomware groups often fails to incorporate theoretical perspectives from business management (for example on companies' growth strategies), negotiation (for example on trust), psychology (for example on leadership qualities), and other fields that help us better understand the dynamics of ransomware. I introduce a novel conceptual framework known as the MOB framework. It is comprised of three crucial dimensions—Modus Operandi, Organizational Structure and Branding and Reputation—which serve as a transformative lens through which scholars and practitioners can analyze the behavior of ransomware groups with significant societal impact. The MOB framework exposes the inner workings of these criminal groups. It highlights the importance of understanding not only how ransomware groups conduct their operations, but also how they structure themselves and build their brand and reputation.

Secondly, I strive to make an empirical contribution by means of applying the MOB framework to one specific cybercrime group. Numerous ransomware groups are operating today. It has become a sprawling ecosystem of people, organizations and transactions. To crystalize the problem, I decided to focus on Conti as a case study, one of the most notorious ransomware groups that was also responsible for the attack against the government of Costa Rica. Between 2019 and 2021, Conti became the largest ransomware group in the world, based on known cryptocurrency transactions and reported incidents. It was responsible for almost half of the known ransomware activity at the time. In 2021, Conti conducted more than 400 successful cyberattacks against major corporations and other organizations.[50] Chainalysis, a company that tracks virtual currency payments, found that the group earned at least $180 million in 2021. Other estimates suggest that Conti raked in at least $1.2 billion in ransom payments in 2021.[51] The Federal Bureau of Investigation (FBI) describes the Conti ransomware as "the costliest strain of ransomware ever documented".[52]

INTRODUCTION

The last aim of this book is to provide insights for policy. Existing books on ransomware frequently provide valuable recommendations for organizations on how to counter evolving ransomware techniques. Yet the scope of the issue has transcended individual organizations. Ransomware has transformed into a societal problem, requiring engagement at both national and international levels. This book seeks to provide policymakers with a comprehensive understanding of the operational dynamics of organized cybercrime, allowing them to make more informed decisions regarding effective countermeasures against the conductors of ransomware.

The Structure and Arguments

This book unfolds through a series of analytical steps, aligned with the objectives of the book. The first chapter sets the stage by examining the key developments in the evolution of ransomware and how these advancements have transformed them into a national security threat. The early 2000s marked the advent of more advanced encryption techniques used by criminal groups, making it feasible for attackers to encrypt victims' data and demand ransom for its release. Despite this, ransomware was not initially the main focus of cybercriminals, who found more lucrative opportunities in the theft of credit card details, passports, and other personal documents. The narrative shifts as we reach the mid-2000s, witnessing a critical transformation in ransomware's deployment. New variants emerged, leveraging botnets to broaden their reach and adopting cryptocurrencies to streamline ransom transactions.

By 2015, the ransomware landscape was further transformed by the emergence of Ransomware as a Service (RaaS). This model significantly lowered the barriers to entry for aspiring cybercriminals by offering user-friendly platforms equipped with tools for creating and managing ransomware campaigns. RaaS platforms not only simplified the technical aspects of launching ransomware attacks but also introduced measures to keep their users—often referred to as 'affiliates'—engaged by deactivating accounts that have not been used for a while. RaaS also marked ransomware's transition into a more openly operated and professionalized venture within the cybercriminal underground. By advertising their services on various online forums, RaaS providers aim to establish trust and attract more affiliates by demonstrating openness and consistency in their operations. The tactic of double extortion—where criminals also threaten to publish stolen data unless a ransom is paid—introduced a new level of threat to the ransomware attacks. Over the past few years, we witnessed the move towards even more professionalized structures among ransomware groups. It has made ransomware a significant threat to human and national security, leading to numerous major incidents. I identify these groups at the helm of this alarming progression in criminal activity, as *ransom war groups*, beyond just ransomware groups.

In the second chapter, the MOB Framework is introduced, aimed at enriching our insight into the dynamics of these groups. Through the use of the framework, I explain how high-end ransomware groups—ransom war groups—are markedly different from high-end nation-state actors, known as APTs. Crucially, these ransom war groups face the challenge of the Ransomware Trust Paradox, which APTs do not. This chapter subsequently explores how, despite their involvement in hacking and data theft, ransomware groups must build trust with their victims.

Chapter three explains how to apply the MOB framework effectively. I highlight that the MOB framework assumes that criminal ransomware groups' modus operandi, organizational structure, and branding are interconnected, impacting each other's evolution and effectiveness. It acknowledges the strategic intent of these groups, their limited decision-making capacity due to organizational inefficiencies, and their adaptability over time. Additionally, the framework considers these groups' awareness of their public perception and rejects the notion that they are driven solely by financial motives, recognizing a blend of economic, ideological, and personal motivations. Furthermore, I discuss that the study of ransomware group presents a significant challenge due to the diverse and extensive data sources required for a comprehensive understanding. This data includes detailed information on their methods of launching attacks, as well as non-technical information like Bitcoin wallets used for financial transactions. Unpacking the organizational structure of ransomware groups necessitates insights into their internal communications and decision-making processes, while understanding their branding strategies involves analyzing their public communications and negotiation tactics.

In conducting the empirical investigation of Conti, as I explain in chapter four, I find myself in the fortunate position of having access to data encompassing all three critical elements: modus operandi, organizational structure, and branding. A pivotal primary resource for this book is the disclosure of leaked chat messages and documents from Conti. In February 2022, Danylo, a Ukrainian security expert, who had long-term access to the Conti servers, exposed a vast trove of files belonging to the group. This leaked information, unprecedented in its breadth, offers a unique insight into the activities of this growing criminal enterprise. Complementing this primary resource, my research also includes interviews with a broad spectrum of experts, including security researchers, ransomware negotiators, and malware developers, insights from cryptocurrency tracing, government communications, and threat intelligence reports.

Chapter five examines the origins of Conti ransomware, starting with Hermes, an encryption tool possibly developed by North Korean developers. Hermes was subsequently used by the Ryuk ransomware group, which adapted and refined its encryption capabilities for their own operations. This chapter then investigates the relationship between Ryuk and Conti. While often referred to as a "successor" or "descendant" of Ryuk, a closer examination reveals a more complex web of interactions. A more accurate description is that Conti emerged as a spin-off from Ryuk, with an overlap in membership and reliance on TrickBot (initially a banking Trojan), indicating a shared resource, leadership and strategy base between the two entities.

Chapter six delves into Conti's modus operandi—or more informally, 'operational playbook'—detailing its evolution and adaptation from previous actions and the tactics of other ransomware groups. This chapter highlights Conti's organizational innovations, such as the formation of an OSINT team and the development of phishing templates, illustrating their methodical approach to maximizing efficiency and impact.

The discussion shifts to Conti's organizational structure in chapter seven. I reveal that Conti organized itself much like a business, with specialized teams for different tasks like market research, negotiation, and software distribution, aiming to work more effectively and grow its operations. It offered a competitive pay and bonuses to attract skilled individuals. The group even followed a regular five-day workweek, mirroring a typical corporate environment.

Still, Conti faced significant internal challenges. Efficiency was often lacking; communication between members frequently broke down, and many were dissatisfied with their jobs. The leader, known as Stern, would disappear for months at a time without any warning to his team. Sometimes employees were not paid, or they too would vanish for long stretches. Disagreements over which targets to pursue were common, showing a lack of unified direction.

Chapter eight details how Conti, mirroring a corporate model, prioritized growth and expansion as its key objectives. The chapter highlights the tactics Conti employs to outpace competitors and shape the future landscape of cybercrime, revealing the group's ambition for broader market reach and deeper penetration. This is demonstrated by their targeted attacks on larger organizations and efforts to establish a global presence. Moreover, Stern, the group's leader, is keen on diversifying Conti's criminal portfolio, venturing into areas like carding, crypto pump and dump schemes, and social media platforms.

Despite these ambitious endeavors, or perhaps because of Stern's dispersed focus, innovation in Conti's core product—the encryption tool known as the 'locker'—has been notably stagnant. The latest version of the locker was less effective than its predecessors. Stern's attempts to diversify and create independent brands separate from Conti frequently backfired, with these endeavors often leading back to the group and failing to detach from the core identity. This situation underscores a critical issue: Conti's organizational structure, which was intended to support innovation and market expansion, actually contributed to operational inefficiencies. Stern's entrepreneurial spirit, while expansive, often resulted in fragmented initiatives that did not enhance the fundamental aspects of Conti's ransomware operations. The misalignment between leadership ambitions and organizational capabilities compromised the effectiveness of their primary 'product'.

Chapter nine covers the complex web of Conti's affiliations, particularly focusing on its interactions with the Russian government. While I dispel the notion of Conti operating under constant guidance or official directives from state authorities, I unveil a nuanced layer of cooperation. It becomes apparent that Conti sometimes undertakes activities they internally describe as 'pioneering.' This involves a form of unpaid labor, where Conti's hacking prowess is leveraged for tasks that align with the interests or objectives of Russian state intelligence.

The cyberattack on Costa Rica stands as Conti's dramatic final maneuver before the group dissolved on May 19, 2022, with all its websites going offline. Chapter ten delves into the cascade of events leading to Conti's dissolution. This chapter first introduces the 'Smokescreen Hypothesis', which posits that the leaks were part of a deliberate strategy by Conti's leadership to mask a deeper reorganization of the group. Amid the turmoil following the leaks and the operations against Costa Rica, it is suggested that Conti was covertly overhauling its structure, improving communication channels, and honing its tactics, all while gradually transitioning to new brands to stay under the radar. This pivot, seen as a masterstroke of deception, aimed to preserve the Conti brand's influence by subtly shifting focus without attracting undue attention.

Conversely, I subsequently propose the 'Jumping off the Sinking Ship Hypothesis' presenting an alternative view. I suggest that Conti's decline was in motion well before the leaks, with the group's direction already fragmenting towards different ventures. In my view, the leaks merely expedited the inevitable collapse, revealing a group frayed by internal strife and lacking unified leadership. The subsequent actions, rather than being part of a calculated plan, emerged from a chaotic attempt to navigate the crisis, leading to a disorganized and piecemeal disbandment of its members.

Conti's collapse represented a transformative event in the cybercrime landscape, reshaping its dynamics and altering the balance of power. The dissolution of the group created a vacuum, prompting employees to seek alliances with other ransomware entities. The second part of chapter ten delves into the aftermath of this dispersion, magnified by the significant leaks of Conti's operational secrets by Danylo. The leak of Conti's internal chats and documentation by Danylo played a crucial role, simultaneously aiding security research by revealing the group's methodologies and internal structure, and inadvertently providing a playbook for other ransomware groups. As a result, the cybercrime ecosystem saw the emergence of new factions: 'Copycats' mimicking Conti's methods without direct ties, 'Offshoots' of former Conti members continuing under new guises, and 'Retaliators' using Conti's tools for mission-oriented attacks.

In the final chapter, I discuss the specific countermeasures against the Conti group but place greater emphasis on the wider measures used to counter ransomware threats. An effective counter-ransomware strategy should disrupt the modus operandi, organizational structures, and branding of these groups. Governments have taken various steps to impede ransomware's operational playbook, including issuing alerts, dismantling infrastructure, distributing decryption keys, and imposing sanctions on cryptocurrency exchanges. Additionally, they have targeted the internal networks of these groups through undercover operations and public indictments. However, the approach to undermining ransomware groups' branding and reputation has been lacking. Existing measures, such as doxing, taking down leak sites, and publishing reports on their failures, are insufficient. I propose further actions, including establishing a code of ethics for ransomware reporting and providing journalist training, to ensure responsible coverage and reduce the glorification and influence of these criminal groups.

1

FROM RANSOMWARE TO
RANSOM WAR GROUPS

In June 1988, the fourth International AIDS Conference convened in Stockholm, Sweden. Distinguished attendees, including senior government representatives, academics, and other leading experts, converged at this event to examine the epidemiology, clinical management, and preventive strategies related to this pressing public health issue.[1]

About a year later, the attendees of the conference received a peculiar mailing: a floppy disk labelled "AIDS Information Introductory Diskette Version 2.0". This same floppy disk founds its way to the subscribers of a prominent London-based magazine, *PC Business World*.[2] The floppy disk appeared to contain an interactive software program called "AIDS Information". The program inquired about the respondents' habits and medical histories, and subsequently calculated the risks associated with contracting AIDS. For those respondents categorized as high-risk, the program spared no words, spitting out a warning message you would now see on the back of cigarette packages: "Your behavior patterns are extremely dangerous and they will very likely kill you".[3]

Unbeknownst to the recipients, the floppy disk also contained a malicious program that infected AUTOEXEC.BAT, a DOS and early Windows system file that runs commands automatically at startup to configure the system environment. The virus did not affect the computer's booting process, but counted the number of times the computer was powered on. Once a specific threshold, typically ninety times, was crossed, the malware sprang into action, encrypting the names of all the files on the main hard drive. The price tag for unlocking the files on the device came in two flavors: a temporary lease for US$189 or a purportedly lifetime lease for US$379. The displayed message, the first ever-recorded ransomware notice to victims, read as follows:

```
Dear Customer:

It is time to pay for your software lease from PC
Cyborg Corporation.

Complete the INVOICE and attachment for the lease
option of your choice.

If you don't use the printed INVOICE, then be sure
to refer to the important reference numbers below in
all correspondence. In return you will receive:

- a renewal software package with easy-to-follow,
  complete instructions;
- an automatic, self-installing diskette that anyone
  can apply in minutes.

Important reference numbers: [….]

The price of 365 use application is US$189. The
price of a lease for the lifetime of your hard disk
is US$378. You must enclose a bankers draft,
cashier's check or international money order payable
to PC CYBORG CORPORATION for the full amount of $189
or $379 with your order. Include your name, company,
address, city, state, country, zip or postal code.
Mail your order to PC Cyborg Corporation, P.O. Box
87-17-44, Panama 7, Panama.

              Press ENTER to continue
```

The programs were written by Joseph L. Popp, who received a doctorate in evolutionary biology and had studied at Harvard University. He spent much of his life living in various African countries.[4] With this message appearing on the screen, Popp presented the computer lock as a voluntary licensing agreement in an attempt to stay within the bounds of the law. He tried to make the case that users of the AIDS program would bear full responsibility for any computer freezes, as his license agreement explicitly warned that failure to make the payment could "adversely affect" the computer.[5]

Popp had set up an elaborate scheme to receive the money. A cheque, bank transfer or international money order had to be sent in an envelope to a post office box in Panama, payable to PC CYBORG CORPORATION. Yet the malware relied on symmetrical encryption, a type of encryption that employed a single secret key for both data encryption and decryption, rendering it relatively easy to decrypt. Soon after the release of the virus, security researcher Jim Bates was the first to produce a reliable removal and retrieval program from the victims.[6] The two programs were distributed free of charge, providing much-needed relief to those impacted.

Only those who panicked and wiped their own data suffered lasting harm. Tragically, a university in Milan permanently erased ten years of astronomical observations, and an AIDS research center in Bologna lost a decade's worth of critical data. None of Popp's victims were known to have made any payments. Ironically, Popp's only income came from the Computer Crime Unit of Scotland Yard's Fraud Squad, who sent the fee during their investigation. PC CYBORG CORPORATION failed to fulfill their claim of providing a decryption tool.

To attribute the AIDS virus to Joseph Popp, Scotland Yard relied on several conventional investigative techniques. PC CYBORG CORPORATION's establishment was traced back to a phone call from Addis Ababa, Ethiopia, where an individual identifying as "Elizabeth Ketema" claimed responsibility. This same person had procured a mailing list from a Nigerian software company for $2,000. The investigators observed a remarkable resemblance in appearance and handwriting between Ketema and Popp. Investigators uncovered additional evidence pointing to Popp as the culprit. Popp's fingerprints were discovered on both the disks and inside several envelopes, each adorned with postmarks from London's Kensington area—an area not far from Popp's residence during that period. Moreover, antigen testing of the saliva from the stamps used to mail the floppy disks aligned with a sample obtained from Popp at a later stage.[7]

After the FBI seized Popp's computer, containing the AIDS program's programming, the British government requested his extradition. Popp claimed that psychiatric medication hindered his understanding of the proceedings, and specialists examined him.[8] Upon his arrival in England, Popp's conduct became progressively bizarre. The *Virus Bulletin*, a computer magazine, wrote that Popp's "recent antics have included wearing a cardboard box, putting hair rollers in his beard to protect himself from 'radiation' and 'micro-organisms' and wearing condoms on his nose".[9] While some psychiatric professionals diagnosed him with severe mental illness, not all concurred with this evaluation. His bizarre conduct in England raised enough concern that he was deemed unfit for trial in November 1991. Eventually, Popp established a butterfly conservatory in Oneonta, New York.[10]

Popp was an early pioneer of ransomware, but significant differences separate his exploits from those of modern-day ransomware groups. Unlike Popp, modern ransomware groups have successfully found ways to monetize their malicious activities. This achievement can be attributed to various innovations, including improved encryption methods and streamlined payment processes, among others. Today's ransomware groups operate with a well-defined modus operandi, thriving within a well-funded and highly professionalized criminal ecosystem, characterized by increased specialization. Ransomware groups have evolved into recognized brands, utilizing their established reputation to engage with the internal criminal sector, victims, and the general public.

This chapter discusses significant milestones in the development of ransomware, and what turned them into a significant threat to human and national security.[11] It starts with the adoption of better encryption techniques by criminals, enabling them to effectively hold data for ransom. The use of botnets subsequently expanded their operational reach, while there was also a shift away from prepaid card systems in favor of cryptocurrencies such as Bitcoin, which provided anonymity and ease of transaction. Following these developments, the emergence of Ransomware as a Service (RaaS) allowed for a better division of tasks within the cybercriminal community, making it easier for newcomers to participate. Tactics evolved further to include double extortion, where attackers threaten to publish stolen data unless a ransom is paid. The final shift saw the professionalization of ransomware groups. It also increased their intent and capability to target major organizations, maximizing their ransom potential. I refer to the ransomware groups at the forefront of this troubling trend in the criminal ecosystem as *ransom war groups*.

Towards Stronger Encryption

The foundational success of any extortion effort lies in eliminating all avenues of escape for the victim. At the core of this use of ransomware is the deployment of strong encryption, which effectively locks victims' data or computer systems, compelling payment for its release, without alternative escape. While the knowledge for implementing strong encryption had been available for quite some time, its widespread adoption by ransomware groups took several years to materialize.

The year 2006 witnessed a significant milestone with the introduction of the Archiveus Trojan. Once a computer fell victim to Archiveus, it copied all files from the user's "My Documents" folder into a single file called EncryptedFiles.als, encrypting them in the process.[12] Archiveus subsequently removed the original files, leaving only the encrypted copy behind. Archiveus was the first to employ the asymmetric encryption method known as RSA. The RSA algorithm, named after its creators Ron Rivest, Adi Shamir, and Leonard Adleman from the Massachusetts Institute of Technology, was publicly described for the first time in 1977.[13] Unlike the symmetric encryption used by Popp, RSA's asymmetric encryption involves one public key and one private key. The public key can be shared openly, while the private key must be kept secret. RSA allows data to be encrypted using either the public or private key, with the opposite key used for decryption. RSA's encryption security lies in the challenge of factoring large integers resulting from two large prime numbers. While multiplying these numbers is simple, factoring them back into the original primes is considered very challenging.

Popp pretended the customer had to pay for a software lease, as a social engineering technique to coerce victims into paying the ransom. Archiveus applied a different technique to apply more pressure on the victim. When victims of Archiveus tried to access their encrypted files, they were directed to a text file with instructions on how to regain access to the data.[14] The instructions falsely accused victims of visiting illegal porn sites, likely to embarrass them and dissuade them from seeking external assistance.[15] The instructions go on to state:

> Do not try to search for a program that encrypted your information—it simply does not exist in your hard disk anymore. Reporting to police about a case will not help you, they do not know the password. Reporting somewhere about our e-mail account will not help you to restore files. Moreover, you and other people will lose contact with us, and consequently, all the encrypted information.[16]

Whereas Popp never provided a decryption tool to its sole payer, the victims from Archiveus did receive a long password from the hackers to decrypt the files after they made a purchase from specified online pharmacies. However, flaws in Archiveus' encryption meant that even with the password, victims often still lost their files.[17]

Around the same period, other cybercriminals were exploring the use of RSA encryption. An example was GPCode, which targeted Windows users through spear-phishing emails. These emails, seemingly from reputable Western companies, offered attachments with enticing details on salary and benefits, sourced from job.ru, a major Russian job portal. The catch wasn't immediate; opening the attachment would install a malware that later downloaded GPCode, effectively masking the initial source of infection and leaving many unaware of the true entry point of the ransomware.

The first instances of GPCode were detected by Kasperky Lab in December 2004. They noticed that GPCode predominantly targeted Russian businesses such as banks, advertising firms and real estate agencies, employing basic encryption techniques. By June 2005, a second outbreak also targeted almost exclusively Russian entities, attempting to deploy a more complex encryption algorithm, which, nonetheless, security experts managed to decrypt with relative ease.[18]

However, in early 2006, a significant shift occurred. It appeared that the creator of GPCode had spent some time studying encryption—possibly drawing inspiration from Archiveus—to release a new variant of GPCode using RSA encryption algorithms.[19] As the ransomware evolved, the encryption key became progressively longer. It began with a 56-bit key, then advanced to 67 bits, followed by a 260-bit RSA key, a 330-bit key, and finally reaching a 660-bit key with the release of version Gpcode.ag on July 4, 2006. This progression in key length significantly complicated the decryption process for those trying to assist the victims.

Despite the greater sophistication of the encryption, the ransom demands remained modest, starting at 2,000 rubles (about $70 at the time) and even decreasing to 500 rubles (approximately $20). The tactic behind Archiveus was not to amass large sums from individual victims but to capitalize on the volume of payments.

Today, an unspoken understanding prevails that ransomware operators refrain from targeting the Commonwealth of Independent States (CIS), a regional organization formed during the dissolution of the Soviet Union with primarily former Soviet Republics. It is the region known for hosting the majority of these attacks.[20] As a consequence, law enforcement in the CIS region tends to ignore their activities. However, GPCode's history demonstrates that this rule was not always in place; it emerged and solidified as the cybercrime community evolved over time.[21]

Botnets and Bitcoin

During the early 2000s, ransomware had not yet become a dominant form of cybercrime, overshadowed by more lucrative activities like trafficking in stolen credit cards, passports and other documents. This era, spanning the late 1990s and early 2000s, coincided with the internet becoming mainstream and the rise of social media and online entertainment platforms. It was during this period that a North American group launched Counterfeit Library in 2000, initially as a space for victims of scams to share their experiences.[22] However, it quickly evolved into a marketplace for counterfeit documents.

Meanwhile, the Russian-speaking cybercriminal community saw the advent of CarderPlanet in 2001, marking a significant milestone in the evolution of cybercrime.[23] CarderPlanet distinguished itself (from Counterfeit Library) by becoming the first global cybercrime marketplace, recognized for its organized structure and adherence to a strict hierarchy, drawing inspiration from the ranks of the Sicilian Mafia—a nod to the influence of *The Godfather*.[24] This platform laid the groundwork for the carding market's expansion over the next decade, leading to the emergence of numerous sites dedicated to these illegal exchanges.

Ransomware only began to carve its niche in the cybercriminal world in the mid-2000s with the introduction of new variants that simplified targeting and opened direct paths to financial exploitation. A significant breakthrough came in 2013 with the appearance of CryptoLocker, which introduced several innovations.[25] CryptoLocker was the first ransomware to rely on a botnet infrastructure, specifically the Gameover Zeus botnet, for distribution. A botnet infrastructure refers to a network of infected computers, controlled remotely to execute coordinated attacks without the users' knowledge. This botnet propagated through spam emails, leading to the download of malware that enabled hackers to perform a range of malicious activities, including disabling system processes, stealing banking information, and installing the CryptoLocker ransomware.

Another key distinction of CryptoLocker was its encryption strategy, employing 2048-bit RSA key pairs—far surpassing the complexity of the 660-bit key used by GPCode and rendering brute-force decryption attempts futile.[26] In certain versions of CryptoLocker, failing to meet the initial ransom deadline allowed victims a second chance to retrieve their files at a significantly higher cost, with ransom amounts fluctuating across different versions and currencies.[27]

The final aspect that makes CryptoLocker stand out is that payments could not only be made with prepaid card systems like Paysafecard and Moneypak, but also by Bitcoin. Initially, prices were established at $100, €100, £100, two Bitcoins, or similar amounts for different currencies. Some reports suggest that 41,928 Bitcoins circulated through four Bitcoin accounts linked to CryptoLocker, translating to over $27 million in payments based on the Bitcoin value at the time.[28] That is over $3 billion in Bitcoin value as of March 2024.

The infrastructure for CryptoLocker's distribution, the Gameover Zeus botnet, was masterminded by Evgeniy Mikhailovich Bogachev. Recognized as a high-value target, the FBI has offered a reward of up to $3 million for information that could lead to Bogachev's arrest or conviction. He was "last known to reside in Anapa, Russia" and "is known to enjoy boating and may travel to locations along the Black Sea in his boat".[29]

A collaborative international effort eventually led to the dismantling of the Gameover Zeus and CryptoLocker operations. In a public statement, U.S. Assistant Attorney General Caldwell stressed the complexity of cybercriminal operations like Gameover Zeus and CryptoLocker, noting the successful disruption of these networks through a collaborative effort involving international and private sector partners:

> These schemes were highly sophisticated and immensely lucrative, and the cyber criminals did not make them easy to reach or disrupt [...]. But under the leadership of the Justice Department, U.S. law enforcement, foreign partners in more than 10 different countries and numerous private sector partners joined together to disrupt both these schemes. Through these court-authorized operations, we have started to repair the damage the cyber criminals have caused over the past few years, we are helping victims regain control of their own computers, and we are protecting future potential victims from attack.[30]

The FBI reported that Gameover Zeus compromised more than 250,000 computers, resulting in losses exceeding $100 million.

Within its mere seven months of operation, CryptoLocker left a profound impact on the cybercrime community. Its success served as undeniable evidence of the great profit potential within this form of cybercrime. Shortly after its emergence, security researchers encountered numerous CryptoLocker clones in the wild, prompting other criminals to join the fray, eager to take part in this lucrative enterprise.[31]

CryptoWall emerged as its most remarkable successor. During the period from mid-March to late August 2014, CryptoWall proliferated extensively via spam phishing emails, infecting a staggering number of over 600,000 computer systems. Its impact extended to the encryption of more than 5.25 billion files. Despite its widespread infection, only a small fraction of victims opted to pay the ransom. As per one report, merely 0.27 percent of victims, amounting to 1,683 individuals, opted to pay the ransom, which averaged around $500, to receive the decryption key.[32]

Cybersecurity company CrowdStrike describes this period as "the true inflection point for ransomware's hockey-stick growth".[33] Developers in the criminal ecosystem established more specialized operations to craft better ransomware code and exploit kit components, flooding the underground hacking marketplaces with their nefarious offerings. With the specialized expertise of developers flowing from other parts of the cybercrime market towards ransomware, it was empowered to further proliferate and evolve into an increasingly formidable threat. This led to another development: Ransomware as a Service.

Ransomware as a Service

In the early 2000s, the emergence of Software as a Service (SaaS) began transforming how businesses operate by simplifying and reducing the cost of using software. Instead of purchasing and managing software on their own systems, companies could now subscribe to software services, eliminating the need for extensive installations and ongoing maintenance. This shift allowed businesses to access software remotely without significant investment in IT infrastructure, opening up access to advanced software solutions once exclusive to large corporations, even for smaller businesses.

By 2015, this SaaS model had evolved into a new variant used by ransomware groups, dubbed Ransomware as a Service (RaaS).[34] RaaS operates on a collaborative framework, delineating roles between the RaaS operator and affiliate. The operator supplies affiliates with the necessary tools, infrastructure, and support for conducting ransomware attacks. This includes recruiting affiliates via online forums, providing them with tailored ransomware packages, and setting up a command-and-control dashboard for campaign monitoring. Operators also manage victim payment portals and aid in ransom negotiations, sometimes extending technical support to ensure the affiliates' operations run smoothly.[35]

The affiliates are the ones executing the ransomware attacks. They gain access to the ransomware by paying the RaaS operator, which can be through a one-time fee, a subscription model, or a profit-sharing scheme based on the ransoms collected. It is common for affiliates to write the ransom notes and handle communications with the victims via chat services. Crucially, they normally manage the decryption keys, which are essential for unlocking the victims' data once the ransom is paid.[36]

RaaS offerings range from basic, cost-effective options to more sophisticated and expensive packages. An example of a low-cost ransomware variant was Stampado. It was offered on the dark web for just $39, providing a lifetime license.[37] A pioneering example of RaaS aiming to minimize effort for affiliates was the Shark Ransomware Project, launched in mid-2016. Setting it apart from the typical ransomware hosted on the anonymous Tor network, Shark was accessible via a publicly available WordPress site. Affiliates simply needed to complete a form specifying their requirements to create customized ransomware. In return for facilitating this streamlined process, Shark's developers took a 20 percent cut of any ransom payments collected.[38]

Investors often emphasize the importance of scalability in business, distinguishing it from mere growth by its ability to increase revenue without corresponding increases in costs. This concept of scalability is crucial for venture capital firms when evaluating startups for potential funding. The advent of RaaS has dramatically demonstrated the power of scalability in the cybercriminal ecosystem. RaaS allows for the centralized development of easy-to-use interfaces and tools, like encryption generators, decryption software, and victim communication and monitoring platforms.[39] This model enables ransomware developers to amplify their reach and impact significantly without proportionately increasing their operational costs, thereby scaling their operations effectively.

Advertisement and Double Extortion

As RaaS emerged, the developers behind ransomware began to engage more openly on underground forums. By showcasing their services prominently, RaaS providers aimed to build trust and attract potential affiliates, relying on the principle that a prominent presence could bolster their credibility.[40] In March 2017, an individual using the alias "polnowz" posted an advertisement for "Fatboy", a new RaaS offering, on a Russian criminal forum. "We invite you to take part in a partnership for the monetization of downloads with help of the Fatboy encryption software. Limited partnership", the advertisement reads.[41] It is explained that purchasers of Fatboy work directly with polnowz, communicating through Jabber, an instant messaging platform, and receive immediate payment upon ransom receipt from victims. Fatboy's pricing strategy was notably based on the Big Mac Index, a concept from *The Economist* used to gauge purchasing power of different currencies, meaning victims in higher cost-of-living regions would face steeper ransom fees.[42] The advertisement detailed other features such as a multilingual interface, automatic decryption post-payment, and a comprehensive partner panel, while also specifying non-operation in the Commonwealth of Independent States.[43,44]

Fatboy's detailed and public advertisement strategy was aimed at rapidly building trust within its criminal customer base. This trend of public, detailed advertisements for ransomware services was further amplified by GandCrab, which transformed ransomware into a media-centric business. GandCrab excelled in branding, marketing, outreach, and public relations, engaging continuously with customers, partners, victims, and security researchers to craft a new type of ransomware enterprise.[45]

Offering a user-friendly RaaS model, GandCrab attracted novices to the field, who, as they became more adept, contributed to the evolution of GandCrab's techniques and eventually launched their own ransomware initiatives. This resulted in a swift development cycle for GandCrab, with the malware undergoing constant updates to bypass security measures.[46] By January 2018, GandCrab had seen at least five major updates, introducing new features and bug fixes that posed significant challenges for the cybersecurity community to mitigate.[47]

An important figure in GandCrab's affiliate program was 'Truniger'. On April 28, 2019, Truniger released files on the hacker forum Exploit that had been extracted from CityComp, a German IT services firm, following a ransomware attack. The post disclosed that the files were made public for free because CityComp had declined to pay the ransom. Although Truniger did not specify the attackers' identity or their affiliated group, the operation is attributed to the 'Snatch team' by German media. Subsequently, the Snatch team published 13 GB of data from an Italian insurance company on the Exploit forum, including sensitive information such as insurance checks, bank transfer details, and personal data of Italian citizens.[48]

Team Snatch was instrumental in popularizing the double extortion tactic, where attackers not only encrypt the victim's data but also threaten its public release. This extortion approach, as demonstrated by incidents like the Lehigh Valley Health Network data breaches, places victims in an exceedingly precarious position, enhancing the attackers' leverage. This evolution in ransomware strategy meant that incidents now often qualify as data breaches, potentially subjecting victims to legal obligations such as notifying affected individuals. Furthermore, it altered the interaction between ransomware operators and the media, with attackers increasingly exploiting public reporting to pressure their victims.

An incident on November 15, 2019 that underscores this shift occurred when Lawrence Abrams, editor of the cybersecurity news site *Bleeping Computer*, received an email from the 'Maze Crew' while finishing his workday. This group is notorious for deceptive spam campaigns pretending to be a government agency and had previously targeted Abrams.[49] Back in May, he and a fellow security researcher had dissected Maze's code, uncovering a reference to *Bleeping Computer*. Subsequently, Maze escalated their provocations by embedding Abrams' email address in malware deployed in a spate of attacks across Italy.[50] The situation intensified with the November email, transforming Abrams into a participant in Maze's operations. The message revealed that Maze had compromised Allied Universal, a major security services company based in Pittsburgh, boasting a workforce of 800,000:

> I am writing to you because we have breached Allied Universal security firm (aus.com), downloaded data and executed Maze ransomware in their network.
>
> They were asked to pay ransom in order to get decryptor and be safe from data leakage, we have also told them that we would write to you about this situation if they dont pay us, because it is a shame for the security firm to get breached and ransomwared.
>
> We gave them time to think until this day, but it seems they abandoned payment process.
>
> I uploaded some files from their network as the data breach proofs. If they dont begin sending requested money until next Friday we will begin releasing on public everything that we have downloaded from their network before running Maze.

The email included a small sample of the purportedly stolen files to substantiate their claims. In the follow-up emails, the Maze Crew explained they demanded 300 Bitcoins, then worth approximately $2.3 million, to decrypt all the files on Allied Universal's computer systems. They explained that the exfiltration of the files, and potential leak, was done to create further leverage to have the victim pay the ransom. Maze also threatened to start a spam campaign using Allied Universal's domain name and email certificates if payment was not made, introducing a new level of extortion tactics. This approach, termed "triple extortion", not only involves demanding a ransom for data decryption (single) and threatening data leakage (double) but also includes the threat of additional attacks if the ransom is not paid.

Following the creation of a leak site by Maze, this multi-extortion method became a trend among ransomware groups, leading to the creation of leak sites, also known as "shaming blogs", by other groups.[51]

Overall, ransomware attacks became not just about encrypting data, but about stealing sensitive information before the encryption. This complicates the recovery process, as merely restoring data from backups does not address the potential release of stolen information. Ransomware groups leverage 'leak sites' to exert pressure on victims through the threat of reputational damage and regulatory issues. This tactic forces victims to respond under pressure, potentially disrupting a measured recovery process.

Professionalization

Following the adoption of double and triple extortion tactics, there was a marked escalation in ransomware attacks globally.[52] This period also saw the professionalization of ransomware operations, exemplified by the rise of the REvil, also known as Sodinokibi. Emerging in April 2019, REvil is often considered the successor to GandCrab, due to similarities in their codes and organizational structures.[53] This professionalization manifests in the greater planning and structured execution of attacks, featuring a clearer division of responsibilities within the ransomware group.[54] They also expanded their reach, targeting even larger and more secure entities and demanding increasingly high ransoms.

Since its foundation, REvil stood out for its adept use of RaaS model. The person behind the moniker "UNKN" or "Unknown" spearheaded efforts to recruit through various underground forums in May 2019, seeking a small group of skilled affiliates. These recruitment posts promised an attractive profit-sharing model, starting at 60 percent and increasing to 70 percent following three successful operations. To enhance the legitimacy of these offers, UNKN placed a significant cryptocurrency deposit, initially around $130,000, which was later increased to $1 million during further recruitment drives.[55] UNKN was pivotal not just in assembling a skilled team but also in shaping REvil's public image, engaging in interviews with media like *The Record* and YouTuber Russian OSINT, disclosing the group's expansion to around sixty affiliates.[56] In July 2021, UNKN mysteriously disappeared which led to "0_neday" taking over.[57]

After successfully breaching and encrypting the data of their targets, REvil would leave behind a ransom note, promptly alerting the victim to the encryption and directing them to purchase a decryption key to regain access. To create a sense of urgency, the note included a timer and a warning that the ransom amount would double once the timer expired.[58] Victims were directed to negotiate over REvil's Tor site, allowing direct communication with the group's representatives. To exert further pressure, REvil sometimes shared snippets of the stolen data during negotiations.[59] Aware that these discussions were publicly accessible, REvil offered the option for private communication as well.[60]

REvil's infamy surged in August 2019 following a significant supply-chain attack through a Managed Service Provider (MSP), affecting 23 Texas municipalities with a ransom demand of $2.5 million—one of the largest ransom payouts at the time. The group did not just rest on their laurels; they refined their tactics and expanded their reach, ensuring their affiliates had the latest in ransomware technology, targeting both Windows and Linux operating systems.[61] Following TeamSnatch's innovation, REvil quickly embraced the double extortion technique, using it by December to threaten the release of stolen data from the CyrusOne incident.[62] That same month, they also hit Travelex, netting a $2.3 million ransom.[63]

In March 2020, taking a cue from Maze's introduction of a leak site, REvil launched their own platform, dubbed the "Happy Blog". The blog was frequently updated with posts that sometimes mimicked official press releases. It later also introduced an eBay-style auction system for selling data belonging to non-compliant victims.[64] Their most high-profile exploitation of double extortion occurred in May 2020 with the theft of 756 GB of data from GSMLaw, a law firm with clients including Donald Trump, Madonna, and Lady Gaga. REvil made headlines by auctioning off this high-profile data after GSMLaw declined to meet their demands.[65]

In 2020, REvil became the leading ransomware variant, focusing its attacks on organizations in North America and Europe, while notably avoiding CIS countries and Syria.[66] In one interview, UNKN also mentioned a particular interest in companies with cyber insurance, viewing them as more inclined to pay ransoms and even contemplating hacking insurance companies to identify prospective targets from their client lists.[67] Despite numerous instances of REvil engaging in re-extortion—demanding additional ransoms from victims who had already paid once, contrary to their promises of deleting the stolen data—this tactic seemingly did little to tarnish REvil's reputation, as many companies continued to meet their demands.[68]

However, REvil's most impactful actions were still on the horizon. In May 2021, they targeted JBS Meatpacking, the largest meat supplier globally, disrupting its operations and eventually extracting an $11 million ransom to restore data.[69] The pinnacle of REvil's activities came in July 2021 with an attack on a Remote Access Software provider through a supply-chain exploit. They exploited a zero-day vulnerability in Kaseya's Virtual Systems Administrator (VSA) software, spreading ransomware via a fake update. This attack affected over 1,500 companies across the world, with REvil demanding an extraordinary $70 million for a universal decryption key, showcasing the extensive reach of their operations.[70]

Ransom War Groups

As the chapter has shown, ransomware groups have evolved significantly over recent years. Groups like REvil, which are at the forefront of this disturbing development in the criminal ecosystem, can best be described as *ransom war groups* rather than simple ransomware groups.[71] These groups have shown both the capability and willingness to orchestrate operations against significant targets driven by the prospect of financial gain. This not only includes major supply-chain attacks, such as REvil's ransomware attack on 1,500 companies through Kaseya. Ransomware attacks on national government institutions are also prevalent, with examples like Conti's disruptive activities in Costa Rica as well as Cuba ransomware group's attack on Montenegro's Department for Public Relations in August 2022, Quantum's attack on the Dominican Agrarian Institute the same month, and RansomHouse's attack on Colombian government ministries in September 2023.[72] Ransom war groups distinguish themselves by their relatively high level of organization and operational planning. Additionally, they cultivate a distinct brand identity and reputation.

Ransom war groups often deliberately execute operations that directly threaten human lives and critical infrastructure, thereby heightening the consequences of non-payment and enhancing their coercive leverage. A stark illustration of this tactic was the ransomware assault by the Russian gang Qilin on prominent London hospitals in June 2024. The attack led King's College and Guy's and St Thomas' trusts—two major acute hospital trusts in London—to postpone 832 surgeries, including critical procedures such as cancer treatments, organ transplants, and heart surgeries, over the course of a week.[73]

At other times, the drive for maximizing profit has led these groups to inadvertently impact human or national security. Ransom war groups often target organizations whose compromise can have broader security implications, which the attackers may not fully appreciate or even wish to avoid, lest it draw significant governmental attention. A contentious example of this is Darkside's ransomware attack on Colonial Pipeline, a company responsible for nearly half the fuel supply for the US East Coast. The breach of Colonial Pipeline's IT systems prompted the company to cease its operations out of concern the attack could spread further. This action triggered a regional emergency declaration by the Federal Motor Carrier Safety Administration on May 9, leading to panic buying and significant fuel shortages across states like North Carolina and Georgia. Subsequently, on May 9, the Federal Motor Carrier Safety Administration issued a regional emergency declaration for eighteen states. It led to panic buying and petrol shortages in several states, such as North Carolina and Georgia. In response, the Biden Administration explored alternative transportation methods for fuel via trucks, trains, and ships. Furthermore, on May 12, President Biden enacted an executive order to implement new cybersecurity standards for software sold to the federal government and to set up an incident review board to extract lessons from major hacking incidents.[74] Yet, Darkside later stated that it had not intended to precipitate such severe societal repercussions and would monitor its targets more closely in the future. The sincerity of such statements is of course debatable given the group's track record and previous PR stunts.

The next two chapters aim to provide a conceptual framework that enhances our understanding of the dynamics within these groups. This structured approach allows us to examine the distinctions between ransom war groups and traditional ransomware groups as well as nation-state actors. Additionally, it facilitates more thorough case study research and supports the development of policy recommendations to counter these criminal actors.

2

THE MOB FRAMEWORK

Over the past decades, the ransomware ecosystem has undergone a remarkable transformation. As I have shown in the previous chapter, what once seemed like sporadic cyber nuisances have evolved into elaborate and organized endeavors. What I termed *ransom war groups* now employ calculated methods and even cultivate a distinct brand identity, making them significant threats to both human and national security.

To better understand this shift in the world of cybercrime, in the coming two chapters I develop the MOB framework. This framework comprises three interrelated elements—Modus Operandi, Organizational Structure, and Branding and Reputation—each open to individual examination. The first, modus operandi, covers the specific methods and tactics used in ransomware attacks, detailing the execution processes. The second element, organizational structure, examines the internal hierarchy, roles, and communication channels within these groups, highlighting how they coordinate and govern their operations. Lastly, the branding and reputation element focuses on the ways these groups manage their public image through communication, negotiation tactics, and overall presentation.

In this chapter, I first use this three-part framework to explain how ransom war groups differ from traditional ransomware groups, building on the historical overview from the preceding chapter. I then compare the differences in modus operandi, organizational structure and branding between ransom war groups and APTs. I explain that ransom war groups and APTs present marked differences, especially in their operational goals and public visibility. While APTs focus on long-term, covert cyber espionage, often state-affiliated and stealthy, ransom war groups seek immediate financial gain. Unlike APTs, these criminal groups face the *Ransomware Trust Paradox*, needing to simultaneously breach trust through attacks while ensuring victims of their reliability for decryption and not leaking the data after ransom payment. The chapter then goes on to explain in more detail how ransom war groups establish this trust.

A Break From the Past

Ransom war groups differ from the ransomware groups of earlier years in several distinct ways. First, gone are the days when ransomware attacks were launched indiscriminately, hoping to snag any vulnerable victim that crossed their path. Ransom war groups conduct their operations with a much higher level of calculation. Big-game hunting entails orchestrating large-scale attacks with substantial ransom demands. Ransomware payloads employ advanced encryption algorithms and techniques that make decryption without the attacker's key nearly impossible. Negotiation processes have become more formalized, involving dedicated communication channels, countdown timers, and threats of data exposure. The rise of cryptocurrency has enabled ransomware groups to conduct global transactions with relative anonymity and greater efficiency.

Furthermore, today, many ransomware groups have embraced a dual tactic of both encryption and extortion, leveraging the threat of data leaks to amplify their impact. By threatening to expose sensitive information, they create a sense of urgency and panic among victims, compelling them to comply with ransom demands. This tactic not only maximizes profits but also adds to the group's reputation as ruthless and credible threat actors.[1]

Second, it is not just the modus operandi that have transformed. What was once the domain of solitary hackers, like Joseph Popp or polnowz, has evolved into a collaborative endeavor akin to specialized company. Ransomware attacks have transcended the image of a lone hacker or disgruntled biologist, tapping away at a keyboard. These attacks are now orchestrated by organized groups that function in teams. Each member has a specific role—be it the developer creating the malicious code, the negotiator liaising with victims, or the money mule funneling illicit profits. Beyond just RaaS, a broader gig economy surrounding ransom war groups has emerged, offering opportunities to outsource a vast array of tasks throughout the attack sequence. This degree of coordination and specialization can sometimes mirror that of small businesses. Yet it is essential not to exaggerate the proficiency of these criminal groups; we must recognize their professionalism but also be aware of their inherent flaws.

Third, perhaps the most striking transformation lies in the realm of branding and reputation. In the past, ransomware attacks were anonymous and transactional—an encrypted file and a demand for payment. Now, the narrative has changed. Ransom war groups have mastered the art of branding their attacks, creating distinct identities that set them apart. They often establish official websites and public-facing portals where victims can make payments and negotiate. They adopt professional-looking logos, user-friendly interfaces, and even offer customer support to guide victims through the payment process.

Ransom war groups must navigate the *Ransomware Trust Paradox* in their operations. Despite engaging in deception by breaching systems and encrypting data, these criminal groups need to assure victims of their reliability and capability—not only promising to refrain from data leaks but also guaranteeing effective decryption upon payment. This assurance is crucial; no organization wants to pay a lot of money and then get nothing in return, or receive a broken tool to unlock their files. Similarly, no organization wants to make a payment only to discover that their data has still been released.

To mitigate these concerns, the brand and reputation of the criminal group becomes paramount, influencing a victim's willingness to assume such risks.[2] For ransom war groups, building a trustworthy brand and reputation is not about ethics; it is a business strategy. Without the victims' trust that paying the ransom will result in the safe return and non-release of their data, the whole business model falls apart.[3]

Distinct from APTs

High-end ransomware groups, or ransom war groups, also differ notably from high-end nation-state actors, known as Advanced Persistent Threats (APTs). APT is a term that dates back to 2008 when Greg Rattray introduced it within the US Air Force to describe emerging threats requiring collaboration with the defense industrial base. The concept of APTs gained widespread recognition in 2013 with the release of the APT1 report by Mandiant, a threat intelligence company which is now a part of Google Cloud. This seminal report shed light on one of the more than twenty APT groups from China that Mandiant was monitoring. It identified APT1 as a unique entity engaged in cyber espionage since 2006, with operations that "resembled" the objectives, capabilities, and resources of the People's Liberation Army (PLA) Unit 61398 of China.

Since then, the APT terminology has become a staple in academic, policy-making, and industry circles, with numerous APTs being identified and named by various threat intelligence agencies. An APT Groups and Operations database, initially spearheaded by security researcher Florian Roth, serves as a repository, aggregating data from diverse cybersecurity reports and enriched by insights from cyber threat intelligence experts.[4] This database documents approximately five hundred APTs discovered over the last decade.[5]

Over time, the APT label has acquired distinct implications, initially reserved for describing the most sophisticated cyber actors, whether state-affiliated or independent. Despite the absence of stringent criteria for classification, many APTs, especially those enumerated in Roth's database, are identified with Chinese espionage efforts, often employing rudimentary tactics without necessarily maintaining persistence. Although the majority of APTs made public are linked to state or state-backed entities, the term has broadened to include a variety of actors, from activist groups to criminal networks, reflecting the diverse spectrum of persistent threats in the digital domain.

Ransom war groups stand apart from APTs across all three elements of the MOB framework. First, the conventional models used to assess the tactics, techniques, and procedures of APTs fall short in capturing the complete modus operandi of these criminal actors. Consider the widely recognized Lockheed Martin Cyber Kill Chain model as an example.[6] This model outlines seven stages of an attack, from initial reconnaissance to achieving the final objective, whether it be data theft or system disruption. However, the tactics employed by these ransomware groups necessitate the inclusion of additional phases such as negotiation, payment processing, decryption, data publication, and the management of cryptocurrency transactions to accurately depict their modus operandi. Through their operational lifecycle, ransom war groups also face unique challenges, particularly in the balance between rapid encryption to secure data and the risk of encryption errors that could jeopardize decryption reliability. This balancing act is critical; swift encryption boosts the chances of a successful ransom demand, but errors can tarnish a ransomware group's reputation and deter victims from paying.[7]

Moreover, ransom war groups differ from APTs in their selection of targets and execution of attacks. Focusing on entities likely to yield high returns, such as major corporations or healthcare providers, these criminal actors deploy targeted phishing campaigns and identify key personnel to maximize the impact of their attacks. While they exhibit a certain level of specificity in choosing their victims, ransom war groups remain fundamentally opportunistic, seeking quick financial gains rather than engaging in prolonged efforts to infiltrate highly secured targets. This approach underscores the tactical divergence between ransomware group's focus on immediate exploitation and APTs' long-term, infiltration objectives.

In particular, especially those APTs affiliated with Western governments, follow a more target-centric approach.[8] Agencies such as the NSA and GCHQ are dedicated to accessing designated networks, conducting thorough analyses to select the most effective tools and tactics for each operation.[9] In contrast to ransom war groups, which may pivot to more accessible targets, APTs often remain steadfast in their pursuit of specific systems. Their commitment to a target is unwavering, persisting through challenges and even employing multiple intermediary breaches as necessary steps to achieve their ultimate goal of infiltrating the intended network.

Persistence, or the capacity to maintain stealthy access to compromised systems over extended periods, further differentiates ransom war groups from APTs.[10] While some criminal actors might dwell undetected within a network to scope out data and backups for several months, this level of persistence is not typical.[11] According to a report from Sophos, ransomware groups usually have around eleven days from breaching a network to being detected—and normally they are spotted because they have encrypted the data on the network.[12] In contrast, agencies like the NSA or GCHQ might covertly operate within a network for years, laying groundwork for significant cyber operations without immediate detection, emphasizing the patience that distinguishes APTs from the more transient presence of criminal groups.

Finally, the nature of the sought-after data also diverges between ransom war groups and APTs' modus-operandi. Ransom war groups are driven by the motive of extortion, targeting information that holds significant *value to the victim* to maximize the likelihood of a paid ransom, exploiting the urgency to protect or recover sensitive data. In contrast, APTs, particularly those associated with national intelligence agencies, focus on gathering information that serves their *own interests*.[13]

* * *

Second, we can examine how organizational structures differ. In my previous book *No Shortcuts: Why States Struggle to Develop a Military Cyber-Force*, I investigated the complex challenges that military cyber commands face. This analysis included a look at the diverse skill sets of the personnel, their operational methodologies, and the foundational infrastructure of these commands.[14] I discuss how the success of a cyber command is intrinsically linked to its staff, ranging from vulnerability analysts, who seek to discover system weaknesses, to developers, operators, testers and system administrators who ensure the continuity of operations. Additionally, I highlight the role of frontline support, which aids in everything from account management to procurement, bolstering the command's functionality.

This organizational blueprint finds a parallel in the structure of ransom war groups, which also exhibit a mix of technical and non-technical roles critical for their functioning. Similar to traditional cyber commands and intelligence agencies, conducting ransomware operations necessitate roles for human resources management, intergroup collaboration, and public relations. They allocate specific responsibilities for recruiting, managing leak sites, content creation, and media liaison within their ranks.

Yet, ransom war groups further diversify their operational roles to include positions like callers and negotiators, each serving a distinct purpose within the group's activities. Negotiators, for example, play a crucial role in interacting with victims post-attack to discuss ransom terms and other conditions, showcasing the specialized and evolving nature of roles within these criminal groups that support their unique operational requirements.

Additionally, there are other significant differences; within ransom war groups, it is common for members of the same organization to be unaware of each other's real identities. This anonymity complicates coordination and collaboration over extended periods.

* * *

Concerning the third facet of the framework, ransom war groups prioritize branding and reputation to a much greater extent compared to APTs. APTs typically carry out covert operations, aiming to remain undetected by their adversaries. In the event of discovery, maintaining a shroud of ambiguity to deny attribution holds significant value for APTs. Conversely, ransom war groups operate differently; while they aim to keep their early activities hidden, they actively seek recognition for their attacks. This quest for self-attribution is vital for building their brand equity. It is essential for the development of brand equity that ransom war groups seek. This alters ransom war groups interact with cybersecurity analysis and media coverage.[15] Unlike APTs, which typically avoid public attention, these criminal groups benefit from and even thrive on such visibility.[16]

APTs using ransomware

Ransomware is increasingly a tool of choice for state cyber actors, with North Korea, Russia, Iran, and China each tailoring its deployment to their specific operational goals and motivations.

One of North Korea's most prominent ransomware attacks was WannaCry, unleashed in May 2017 and attributed to the Reconnaissance General Bureau of the General Staff Department, a clandestine operations unit also known as Lazarus Group or Hidden Cobra.[17] This operation targeted Microsoft Windows systems, encrypting data on hundreds of thousands of computers worldwide and demanding Bitcoin payments for decryption keys.[18] Despite the vast reach of WannaCry, it was also "poorly run, shoddily coded, and barely profitable".[19] The malware required manual verification to release the decryption keys, a cumbersome process that limited its profitability at scale. Moreover, relying on only a few Bitcoin wallets made its financial channels easy to track and disrupt, drawing significant law enforcement attention but yielding minimal returns. Craig Williams from Cisco's Talos team described it as a "catastrophic failure" from a ransom perspective, noting its "high damage, very high publicity, [and] very high law-enforcement visibility," but "probably the lowest profit margin" seen among ransomware campaigns of its size.[20]

In recent years, however, North Korean groups have refined their tactics, drawing more directly from organized cybercriminal methods discussed in Chapter one. One example is DEV-0530, a group tracked by Microsoft and self-styled as H0lyGh0st, which has conducted ransomware campaigns since mid-2021.[21] Emulating the double-extortion tactics pioneered by Maze, DEV-0530 pressures victims by threatening to leak sensitive data publicly or directly to clients. DEV-0530's activities overlap with another

North Korean group known as Andariel or DarkSeoul, infamous for the Maui ransomware.[22]

Russia, by contrast, frequently uses ransomware as a tool of disruption rather than profit. A key example of this is NotPetya, which followed WannaCry. In this 2017 operation, the Russian GRU's Unit 74455 gained access to the update servers of Linkos Group, a small Ukrainian software provider, and inserted backdoors into systems of global clients.[23] The backdoors were then used to deploy NotPetya, malware that mimicked ransomware but was designed solely for destruction. Unlike its predecessor Petya, NotPetya had no decryption functionality, making ransom payments irrelevant.[24] Once released, NotPetya spread uncontrollably, wreaking havoc across major industries worldwide, affecting organizations such as Maersk, Saint-Gobain, Mondelez, Reckitt Benckiser, and TNT Express. Even Russian state-owned Rosneft was impacted, illustrating the collateral damage of NotPetya's indiscriminate spread.[25]

Following its further invasion of Ukraine, Russia intensified its ransomware deployments. In one striking case, GRU Unit 74455 deployed Prestige ransomware against transportation and logistics firms in Ukraine and Poland,[26] likely to disrupt critical supply routes supporting Ukraine's defense.[27]

Iran's ransomware activities have increasingly focused on the U.S., often targeting critical sectors like finance, healthcare, and education. An August 2024 Cybersecurity Advisory from the FBI, CISA, and DC3 highlighted that Iran-based cyber operatives commonly establish initial network access and then collaborate with ransomware affiliates to conduct attacks.[28]

In the Middle East, however, Iran's ransomware tactics are most visible, especially in campaigns against Israel. For instance, shortly after the Hamas attacks on Israel in October 2023, the IRGC's Shahid Kaveh Group launched a cyber campaign against Israeli security cameras using customized ransomware, under the alias "Soldiers of Solomon." A Microsoft report revealed that the attackers falsely claimed to have compromised data at Nevatim Air Force Base; however, the footage they posted actually came from a street north of Tel Aviv that also happened to be named Nevatim.[29]

Finally, Chinese cyber espionage groups have occasionally used ransomware, likely as a means of maintaining plausible deniability.[30] In late 2022, media reported on ransomware attacks on Brazil's Presidential Office and the All India Institute of Medical Sciences (AIIMS), one of India's leading healthcare institutions. However, researchers Aleksandar Milenkoski and Julian-Ferdinand Vögele found that ChamelGang, a Chinese APT group, was likely responsible for these attacks, using their CatB file locker tool.[31]

Overall, the use of ransomware by state actors often diverges in purpose and technique from the financially motivated operations of typical ransomware groups. For these state actors, payments are often irrelevant, bypassing the usual challenges of maintaining credibility or managing victim relationships. Without the need for trusted "brands," states employ ransomware primarily to advance strategic goals: creating confusion, disrupting adversaries, or accelerating the pace of operations without the logistical overhead of payment negotiation.

However, North Korea remains a clear outlier, using ransomware primarily for financial extortion. Yet, efforts to mimic criminal ransomware groups have often fallen short. North Korean operators struggle with the complexities of the Ransomware Trust Paradox, finding it challenging to build a recognizable, reliable brand. For example, the initial absence of a ransom note in the Maui ransomware campaign highlighted gaps in North Korea's credibility as an extortionist. Additionally, North Korean operators have attempted to imitate well-known groups like REvil to establish credibility, yet these imitations are frequently recognized by victims and researchers alike, underscoring the challenges North Korea faces in establishing a trusted identity within the ransomware ecosystem.[32]

How Ransomware Groups Establish Trust

If successful ransomware operations hinge on trust, it is important to explore in greater depth how ransomware groups can develop this trust.[33] According to negotiation and business literature, there are three foundational types of trust underpinning relationships: identification-based, knowledge-based, and deterrence-based.[34] The most successful ransomware groups primarily use the latter two types to navigate the *Ransomware Trust Paradox*.

Identification-based trust represents the highest level of trust, where one entity fully understands the other's preferences. A common example of this is a tribe, where members often show greater trust towards fellow tribesmen than outsiders, attributed to common experiences and values.[35] Shared objectives, closeness and similar values foster trust in such environments.

For any ransomware group, establishing such profound trust with their victims is nearly impossible, despite attempts. For instance, the LostTrust ransomware group claims to be "specialists in the field of network security with at least 15 years of experience", who have turned to ransomware due to poor pay for their legitimate hacking services.[36] They aim to project an image of professionalism and shared experience to cultivate trust.

Second, there is deterrence-based trust, which arises in scenarios where the potential costs of ending a relationship or the threat of retaliation outweigh the (short-term) benefits of acting deceitfully.[37] Nobel laureate Thomas Schelling demonstrated that repeated interactions between two parties can establish a pattern of expected behavior. Repeated interactions make it less likely for parties to deceive each other in a single transaction if future, beneficial transactions are anticipated.[38] Another mechanism facilitating deterrence-based trust is known as 'hostage taking'. During the Middle Ages, a lord might take another lord's only son or daughter as a means of ensuring trustworthiness. The potential loss of a valued child often sufficed to ensure that the lord adhered to agreements.[39] Similarly, in today's context, consider a well-known online retailer recognized for their distinctive and high-quality products. Such a retailer would steer clear of deceiving a customer, as they would be concerned that negative feedback could damage their online reputation and dissuade potential buyers.[40]

Ransomware groups' interactions with victims are typically singular events rather than evolving relationships with repeated interactions. Unlike businesses that build relationships over time and create a foundation of trust through repeated positive interactions, ransomware groups typically have a one-time interaction with their victims. This is not a scenario where victims are "ransomwared" continuously by the same group, building a pattern over the years.

However, there is a major role for ransomware recovery companies, negotiators and insurance companies in this landscape. These entities often find themselves dealing with the same ransomware groups repeatedly on behalf of different victims.[41] This repeated interaction can influence the dynamics between these intermediaries and the ransomware operators, potentially impacting the negotiation processes and outcomes.[42]

Trust through reputational hostage taking is paramount for a ransomware groups—hence, branding and reputation is the third facet of the MOB framework. Each time they interact with a victim, they are negotiating not just for a ransom but also for their reputation. Their reputation acts as a "hostage". If they fail to uphold their end of the bargain, they risk damaging their reputation, which deters future potential victims from trusting and engaging with them. Some criminal groups are very explicit about this. For example, Darkside states in their ransom note: "We value our reputation. If we do not do our work and liabilities, nobody will pay us. This is not in our interests. All our decryption software is perfectly tested and will decrypt your data. We will also provide support in case of problems. We guarantee to decrypt one file for free. Go to the site and contact us".[43] Similarly, Karma writes in their message to victims: "Decryption is only possible with a private key that only we possess. Our group's only aim is to financially benefit from our brief acquaintance, this is a guarantee that we will do what we promise. Scamming is just bad for business in this line of work".[44]

Some groups even issue "press releases" to correct mistakes made by journalists that might affect their reputation. One example is the Snatch group. In one of their releases, they state:

> First of all we have nothing to do with the Snatch ransomware project that appeared in 2019 and existed for about 2 years. We

are the Security Notification Attachment (SNAtch for short) Team, a group specializing exclusively in leaked sensitive data. We don't deal with locking companies or critical infrastructure, we don't aim to stop a company from operating by attacking it with software that blocks the control servers. If journalists analyze our work carefully, they will see that not a single client of ours has been attacked by a malware that can be called Snatch. [...] So the main thing that we want to say and convey to you is that the Security Notification Attachment Team (SNAtch for short) has nothing to do with the Snatch ransomware project.[45]

Another example is BlackCat, also known as ALPHV, which emerged in late 2021. They released a 1,300-word article on their leak site titled "Statement on MGM Resorts International: Setting the Record Straight". In this article, they rebuked several publications, including *The Financial Times* and *TechCrunch*, for failing to verify sources and disseminating inaccurate information.[46]

Third, knowledge-based trust is built on behavioral predictability. This type of trust develops as parties come to understand each other's behavior patterns, values, and intentions over time. Several methods can improve understanding and predictability between parties.[47] Regular communication is one such method: for example, trust between teammates can be enhanced by routine meetings and status updates. Another way to bolster knowledge-based trust is through what is often referred to as 'courtship', which revolves around consistent behavior from a potential partner.[48] This approach is vital in situations like partnerships between startups and investors, where early interactions that display reliable and predictable behaviors can lay a strong foundation for trust. This predictability allows parties to better anticipate each other's actions and commitments.

The more established groups have a communication plan to project a sense of behavioral predictability to their victims. Many ransomware actors, such as Cl0p which emerged in early 2019, establish dedicated channels for victims to negotiate ransom payments and receive instructions. They adopt a service-oriented tone, referring to victims as "customers" and themselves as "support", creating the illusion of a legitimate business transaction.

Groups like 8Base, which appeared in 2022, even include FAQs and guidelines on their darknet sites to seem more transparent in their operations.[49]

After receiving the ransom payment, some groups offer comprehensive technical support to assist victims in decrypting their data. This support often includes detailed, step-by-step instructions, troubleshooting services for issues with the decryption key, and occasionally, help desks staffed with operators prepared to assist with the decryption process. To demonstrate their capability and build trust, some groups even decrypt a few files for free, reassuring victims that they have the means and will indeed unlock the files once payment is made. An example of this approach is illustrated in the initial message from the ransomware group HelloKitty to their victims:

> Unfortunately, your files were encrypted, and more than 200 GB of your critical date was leaked from your File, DEV and SQL servers (Administration and Finance, Direzione, Legal, HR, Risorse umane). For a more detailed list of documents, please contact us and we will send you the samples we have. We are also ready to help you recover your files, prevent the spread of leaks, as well as help solve problems in your IT infrastructure that were the cause of the current situation, so that this does not happen again in the future. Just contact support using the following methods and we will decrypt one non-important file for free to convince you of our honesty.[50]

In the same way, Chilelocker ransomware clearly outlines the consequences and options for victims in their communications:

```
YOUR OPTIONS:

---> IF NO CONTACT OR DEAL MADE IN 3 DAYS:

Decryption key will be deleted permanently and
recovery will be impossible.

All your Data will be Published and/or Sold to any
third parties

Information regarding vulnerabilities of your net-
work also can be published and/or shared
```

```
---> IF WE MAKE A DEAL:

We will provide you with the Decryption Key and
Manual how-to-use.

We will remove all your files from our file-storage
with proof of Deletion

We guarantee to avoid sharing any details with
third-parties

We will provide you the penetration report and list
of security-recommendations⁵¹
```

Reflecting on the importance of reputation in earlier discussions, it can be detrimental for a ransomware group if a Google search reveals numerous articles indicating that the group fails to fulfill its promises, suggesting a lack of predictability. Conversely, it is advantageous for them if search results highlight successful data recoveries or portray them as reliable in returning data upon payment.

Brand Strategy of Ransomware Groups

Ransomware groups often come up with meaningful names that helps in their branding. REvil, an acronym for "Ransomware Evil", draws inspiration from the *Resident Evil* film franchise.[52] Meanwhile, the Cl0p ransomware, which made its debut in February 2019, takes its name from the Russian word "*klop*", signifying a bed bug.

Beyond names, these criminal groups create distinctive brand identities through symbols and logos. The Vice Society, a group targeting educational institutions since 2021, uses imagery reminiscent of the video game *Grand Theft Auto: Vice City*. Lockbit, another well-known group, employs a retro-themed logo in red, white and black, ensuring that it features prominently across their communication channels. They even incentivized individuals to tattoo their logo.[53]

This approach to branding marks a departure from the behavior of earlier ransomware like Archiveus, which would disguise its malicious nature under the guise of authoritative entities, misleading victims with warnings of illegal online behavior and demanding fines for release. The practice of naming these variants often fell to external observers, as seen with Jigsaw ransomware, initially dubbed "BitcoinBlackmailer" but later renamed due to its association with imagery from the *Saw* movie series. Early forms of ransomware targeted victims indiscriminately, relying less on a public image or reputation to attract collaborators, given their smaller scale and less sophisticated organization. The emphasis on branding was less pronounced among these predecessors, who relied more on a 'spray and pray' approach targeting smaller victims. Furthermore, these early ransomware groups were smaller organizations that did not rely on external partnerships and affiliates, diminishing the necessity for a well-crafted reputation to lure collaborators or new members.

An important outcome of more developed ransomware groups handling their own branding is the potential to establish multiple brands within the same organizational structure. An analogy can be drawn from The Coca-Cola Company, which offers over 200 brands ranging from well-known soft drinks like Coca-Cola, Sprite, and Fanta, to products like Dasani water, Vitaminwater, and Costa coffee and tea.

There are various benefits to creating multiple brands. First, given that law enforcement and regulatory bodies are constantly on the lookout to disrupt the actions of ransomware groups, the ability to operate under various brands can safeguard against disruptions. Authorities often alert the public to the activities of specific cybercriminal groups, urging non-compliance and actively working to dismantle their networks or expose their encryption keys. By maintaining multiple brands, each with its distinct technology and methodologies, a ransomware organization can ensure the continuity of its operations, even if one brand comes under scrutiny or is compromised.

Additionally, the cybercriminal ecosystem's competitive nature further justifies the need for multiple brands. In a domain characterized by intense rivalries, introducing a new brand can be a move to attract affiliates and operators seeking new opportunities. Early association with an emerging brand may open doors to deeper access within the group, presenting what could be viewed as a progressive step in their criminal "career". Moreover, this approach could attract individuals who are hesitant to work for certain exist-ing brands due to past negative encounters or unfavorable experi-ences. Opting for involvement with a new or up-and-coming brand offers them an opportunity to distance themselves from any prior associations and start anew. In essence, this strategy not only capi-talizes on competition dynamics but also taps into the aspirational aspect of affiliating with a brand that embodies potential success and a chance for better outcomes.

Another significant motive for cultivating diverse brands is the tactical re-engagement with previous victims while preserving the credibility of the organization's other brands. Ransomware actors typically pledge that by complying with the ransom demand, not only will they decrypt the victim's data and refrain from leaking stolen information, but they also commit to avoid-ing any future targeting of the same entity. They often promise that compliance with ransom demands will result in data decryp-tion and a guarantee against future attacks. However, with mul-tiple brands at their disposal, these groups can revisit previous targets under a new guise. Armed with detailed knowledge of the target's systems and valuable data, they can initiate a second round of extortion. Presenting the victim with a new leak site, also known as "shame site", adorned with a different name and logo, possibly using alternative encryption tools, subtly enhances the odds of a successful extortion, all while maintaining the facade of a trustworthy entity in the broader public sphere.

The Need for a New Perspective

Historically, discussions on cyber conflict have primarily centered on the involvement of state-sponsored or affiliated groups.[54] While cybercriminals have been using digital tools for quite some time, their recent prominence demands a shift in our attention. These groups now endanger the safety of citizens as well as undermine international security and stability. Despite their significant impact, the field of International Relations and Security Studies—the aca-demic discipline in which I originally specialized—has paid hardly any attention to these groups. The focus has primarily remained on threats at the state level.[55]

As we explore the expanding influence of criminal ransomware groups—and especially the high-end ransom war groups—it becomes evident that a change in perspective is necessary. This is where the MOB framework becomes crucial. This analytical framework enables us to move beyond the usual discussions and thoroughly analyze the operations of these groups.[56]

3

THE MOB FRAMEWORK IN
ANALYTICAL PRACTICE

The previous chapter is a call to scholars, practitioners, and policymakers to recognize the significance of ransom war groups and
equip themselves with the necessary tools to analyze, anticipate,
and mitigate their actions. The MOB framework emphasizes the
importance of understanding not only how these groups technically
conduct their operations but also how they organize themselves
and project their brand and reputation within the cyber realm.

This brief chapter further explores the application of the MOB
framework as a foundation for more in-depth empirical research
on ransomware, including my detailed case study on Conti. I have
divided this chapter into three parts. First, I explain that the MOB
framework relies on several key assumptions. It emphasizes the
need to study not only their operational playbook but also their
organizational dynamics, branding, and reputation management. It
assumes that these groups have strategic intent but acknowledges
their limitations in making rational decisions due to cognitive and
organizational inefficiencies. The framework posits that Modus
Operandi, Organizational Structure, and Branding are interconnected and adaptable. Additionally, it recognizes that these groups
are aware of how they are perceived by others and may strategi-
cally adjust their behavior, although various factors often hinder

their ability to quickly adapt to changes in law enforcement, security measures, and victim behaviors.

Second, I explain that underlying each element of the MOB framework—Modus Operandi, Organizational Structure, and Branding and Reputation—there are key components that capture the complexities of ransomware groups. These include the techniques and target selection in modus operandi; internal hierarchy and communication in organizational structure; and the group's identity and external interactions in branding and reputation, including possible connections to state actors.

Lastly, I discuss the data requirements to study ransomware groups using the MOB framework. Researchers need diverse data sources, including technical details on tactics, network tools, Bitcoin addresses, communication logs and leaked documents, in order to understand their structure, branding, and reputation. Collecting this data is challenging due to the secretive nature of ransomware groups and ethical considerations, requiring specialized tools and techniques to extract valuable insights while balancing privacy and legal concerns.

Assumptions of the MOB Framework

The MOB framework is based on several key assumptions that underpin its analytical approach to understand these criminal actors. The principal assumption is that a comprehensive understanding of a criminal ransomware group requires studying not only their technical components—which have received the lion's share of attention—but also their organizational dynamics, branding, and reputation management.

The study of ransomware groups—and specifically ransom war groups—through the MOB framework also assumes strategic intent: these criminal ransomware groups make deliberate choices in how they target victims, execute attacks, negotiate with victims, and manage their reputation. However, this does not mean that they are rational, nor that groups are always able to make decisions that provide them with the highest amount of utility. In fact, the organizational structure element strongly suggests that this is not achievable, not only because of the inherent limits to humans' cognitive, decision-making capacity, but also because of organizational inefficiencies that are introduced.[1]

The third assumption of the MOB framework posits that Modus Operandi, Organizational Structure, and Branding are interconnected and influence each other. For example, changes in group structure, such as key members departing or joining, could lead to shifts in the modus operandi of the group. The addition of skilled developers (joining from a different ransomware group) might prompt an expansion of capabilities to encompass new devices—possibly transitioning from solely targeting Windows devices to including MacOS systems for Apple desktops and laptops. When an organization disbands and individuals leave, they might regroup elsewhere, creating a new brand identity. Similarly, another situation could arise where a ransomware group maintains a prominent reputation, yet its organizational structure has significantly dwindled.

The fourth assumption closely tied to this is that ransomware groups exhibit adaptability and the potential for evolution over time. The framework accommodates various possibilities for change and evolution, where specific elements of the MOB framework might transform while others remain relatively constant. For instance, we occasionally witness shifts in the modus operandi while the organizational structure and branding retain a high degree of similarity. Lockbit serves as a fitting example of this phenomenon. Despite the shift in its modus operandi since 2020, along with changes to its affiliate model, LockBit retained its brand identity throughout the years.[2]

Fifth, the framework operates under the premise that criminal groups possess a certain level of awareness regarding how they are perceived by various parties, including their victims, the security community, law enforcement, and potential recruits. This awareness might prompt them to manipulate their behavior in order to influence external perceptions. For example, Russian ransomware groups may adopt a foreign name to avoid the stigma and financial sanctions typically directed at Russian cybercriminal operations. This seems to have been the situation with Yanlouwang, a group that has been operating since late 2021. The group is known for targeting businesses in sectors like information and communication technology, engineering and finance. Yanlouwang gets its name from a Chinese mythological figure, which suggests that the group might have Chinese origins. However, in late October 2022, a Twitter account called @yanlouwangleaks leaked chat logs from the group.[3] The leaked chat logs showed that the group consisted of approximately two dozen people who spoke Russian (members did not know each other's real-life identities).[4]

While the framework acknowledges the adaptability of ransomware groups, it refrains from making a blanket assumption of their agility. The reality is that various human, organizational, and cultural factors often impede ransomware groups from swiftly adjusting their tactics, techniques, and procedures in response to changes in law enforcement strategies, security measures, and shifting victim behaviors.

Key Elements of the MOB Framework

The MOB framework is straightforward, with just three elements. Underlying each element, however, is a greater set of components, which help to capture the finer nuances and complexities of ransomware groups.

To thoroughly examine a group's modus operandi using the MOB framework, it is crucial to address several key aspects. Firstly, there is a need to explore the tactics and tools used by the ransomware group, from initial reconnaissance to data exfiltration and payment. It requires the answering of questions, such as: How is the ransomware group gaining access? Do they often use zero-day vulnerabilities to escalate privileges?[5] Are they predominantly using open-source tools, and are they also developing their own tools?

Secondly, understanding the group's target selection process is essential. This involves identifying preferred industries, geographic regions, or specific types of organizations. For instance, does the group exclusively target Spanish-speaking entities? Do they refrain from targeting entities in Russia and other CIS countries? How aggressively do they pursue targets such as healthcare institutions?

Table 1: Components of the MOB Framework

Component	Definition	Key Elements
Modus Operandi	How ransomware attacks are executed and carried out.	• Techniques and tools employed • Target selection • Response to external entities
Organizational Structure	The internal hierarchy, roles, and communication of the group.	• Hierarchy of roles • Internal communication • Recruitment and retention • OPSEC
Branding and Reputation	The way the group communicates, negotiates, and presents itself.	• Name and identity • External communication • Ethical stance

Lastly, it becomes important to examine the group's responses to external factors, like law enforcement actions. This could involve public statements, shifts in tactics, or even retaliatory actions against law enforcement targets.

To understand the organizational structure, one can equally look at various elements. First, the hierarchy of roles within the group, ranging from leaders to operatives. This involves figuring out members' responsibilities, decision-making authority, and how tasks are divided. In much of the reporting there is a tendency to generalize about the structure of ransomware groups; for instance, it is often said that these groups depend on initial access brokers or follow a RaaS model. However, the reality shows considerable variation among different groups. Some, for example BlackCat, employ an affiliate model and actively recruit individuals on underground forums.[6] This approach allowed BlackCat to scale operations quickly, targeting entities such as OilTanking, a German fuel company, and Swissport, an aviation services company.[7] Conversely, other groups prefer a tighter grip on their operations to maintain control. Royal, a ransomware group that surfaced in January 2022, consists of seasoned individuals who keep their code and infrastructure private. Working in a smaller group, Royal has been more selective with its targets, focusing on high-value corporate entities to secure larger ransoms.[8]

Second, we can study how information flows within a ransomware organization, and between the group and the victims. This includes identifying the tools and platforms used for communication, coordination, and sharing of operational details. Telegram has become one of the platforms where cybercriminals communicate with victims and release data.[9] It is recognized for its safety and encryption policies, though it should be noted that by default, Telegram only encrypts messages during their transit between a user device and its servers, with an option for users to enable end-to-end encryption manually. Equally, there remains variation among groups, with some of the more advanced ones setting up separate web-based chat portals for victim communication.[10]

The third important aspect is how new members are recruited into the group. This requires an investigation of the process of onboarding, including training and integration into the operations. There are also important elements of geographical distribution, whether the group operates from a single location or has members distributed across different regions or countries; and affiliate relations, whether the group collaborates with other criminal organizations, sharing resources, tools, or tactics. One prominent development in this respect occurred in 2021, when the two major hacking forums—XSS and Exploit—announced that they were banning ransomware sales, ransomware rental, and ransomware affiliate programs on their platform.[11] This meant that ransomware groups were compelled to find alternative recruitment channels. A final relevant element to research is the measures taken by the group to promote OPSEC and protect their own members and operations from exposure or infiltration by law enforcement.

Key elements to study within branding and reputation include the group's name and identity, as well as the language and tone used in the group's communications with victims, media outlets, and other cybercriminals. An interesting element here is the interaction between ransomware groups. This can include partnerships, collaborations and conflicts, which can influence how they are perceived within the community. An equally important element concerns a ransomware group's ethical stance within the cybercriminal ecosystem. This includes examining debates about targeting critical infrastructure, healthcare organizations, or other sensitive sectors.[12]

Consider, for instance, the IBM Security X-Force's analysis of approximately forty-five negotiations of victims with Egregor, a ransomware group that was active from September 2020 to February 2021 before being dismantled by an international operation.[13] In one negotiation, an Egregor call center employee explained how they calculate ransom demands, aiming for 5 percent to 10 percent of a victim's potential financial losses.[14] When victims said that they could not pay, negotiators sometimes asked for tax documents. Additionally, Egregor sometimes showed empathy: they offered to decrypt a charity's systems for free, but with the condition that the charity publicize this act to enhance the group's public image. "You will cover in the media the fact that we gave you the decryptors for free due to our social responsibility", Egregor's chat support demanded.[15]

Another revealing example involves the Darkside ransomware group. In October 2020, in an apparent attempt to be perceived as Robin Hood-like characters, the criminals posted receipts for $10,000 in Bitcoin donations to two charities, the Water Project and Children International.[16] "No matter how bad you think our work is, we are pleased to know that we helped changed someone's life. Today we sended the first donations", the group wrote. This is the same group that seven months later was responsible for the ransomware attack on Colonial Pipeline, setting off a chain of disruptive events. They subsequently posted an apology on their dark web site, potentially to reduce heat from law enforcement agencies:

> We are apolitical, we do not participate in geopolitics, do not need to tie us with a defined government and look for other our motives. Our goal is to make money and not creating problems for society.

From today, we introduce moderation and check each company that our partners want to encrypt to avoid social consequences in the future.[17]

Finally, an aspect that cuts across all components of the MOB framework concerns a ransomware group's potential connections to state actors. While traditional branding elements focus on the group's public image, messaging and tactics, the potential alignment with a state introduces a layer of complexity that can significantly impact the group's reputation and operational behavior. It emphasizes the need to explore whether the group tailors its messaging, motivations and goals based on its perceived alignment with a state. Moreover, it extends to how law enforcement and cybersecurity threat intelligence firms respond, considering diplomatic considerations and legal actions against both the ransomware group and its supposed state sponsor.

Data Requirements

A significant challenge that comes with the study of ransomware groups is the considerable need for diverse and extensive data sources. Researchers require detailed technical information to grasp the modus operandi of ransomware groups. This encompasses information about access methods, network tools, and the type of locker used. Additionally, non-technical aspects like Bitcoin addresses for money transfers are also crucial for understanding the playbook of these actors.

Examining the organizational structure necessitates data that reveals the inner mechanisms of ransomware groups. This may include communication logs, chat transcripts, training documents, and forum discussions where group members interact. Information about roles, responsibilities, affiliations and communication patterns can provide insights into the decision-making process, power distribution, and coordination mechanisms within the group.

Finally, understanding the branding aspects requires data on the communication tactics and negotiation strategies employed by ransomware groups. This entails studying ransom notes, negotiation emails, leaked data excerpts, and the history of communications with victims.

Many of these data sources are challenging to gather. Not least, ransomware groups are not inclined to willingly disclose their internal structure or unveil their confidential information. However, ransomware groups often have their internal communications and actions revealed due to breaches in operational security (OPSEC). In the second part of this book, we will specifically explore the leaks related to the Conti group as a detailed case study. But Conti is not an isolated case; other groups such as Yanluowang have also had their internal chat logs leaked online.[18]

Marcello Ienca and Effy Vayena have identified several ethical considerations for conducting research with this type of hacked data: demonstrating unique value, assessing risk-benefit, obtaining consent, ensuring traceability, and preserving privacy.[19] These guidelines ensure that researchers use such data responsibly, particularly when it cannot be obtained by conventional means, and that the benefits of research significantly outweigh potential harms. Based on these criteria, using leaked chats from cybercriminal ransomware groups for research can be ethically justified. These communications offer a rare glimpse into the operations of organizations that significantly harm society. Understanding the internal dynamics of such groups provides crucial insights that can enhance measures to counteract their activities. Furthermore, the anonymity typical within these groups means that the data is generally not personally identifiable, as members often use pseudonyms. This reduces the ethical concerns associated with privacy and consent, making the case stronger for the ethical use of such data in research aimed at public safety and security enhancement.[20]

* * *

Given its importance as a data source, delving into why ransomware groups are often susceptible to operational security breaches—seemingly even more than APTs—is useful. First, the process of integrating new operatives, whom they have not physically met and whose identities are not fully verified, poses a significant challenge. The difficulty of conducting in-depth background

Table 2: Data Requirements of the MOB Framework

Element	Data requirements	Potential sources
Modus Operandi	• Tactics, techniques and procedures • Transaction data and wallet addresses	• Malware samples and C2 infrastructure details. If not collected by the researcher directly, they are commonly found in (commercial) threat intelligence reports • Criminal training manuals • Data from cryptocurrency analysis platforms
Organizational Structure	• Communication logs and chat transcript • Leaked documents and forum discussions • Affiliations and member interactions	• Forum and marketplace data where ransomware group members interact— eg. discussing recruitment, job roles, partnerships, and collaborative activities • Chat logs obtained from secure messaging platforms used by the group's members • Insider interviews • Indictments and other government reports

| Branding and Reputation | • Ransom notes and negotiation emails
• Communication history with victims
• Leak site data | • Ransom notes and communication exchanged between the ransomware group and victims to understand their communication strategies, demands, and negotiation tactics
• Online presence data (eg. 'leak sites') where the ransomware group interacts with victims, shares updates, and posts stolen data
• Victim testimonies |

checks complicates the vetting process, making it arduous, if not unfeasible, to ensure the trustworthiness of new entrants. For those with an established history, decisions might be based on their known track record and endorsements from reliable connections. Yet, for novices within the group, the approach is more nuanced, necessitating a careful and gradual acclimatization to the group's activities and hierarchy.

Although such arrangements may work effectively in smaller, more cohesive units where members have a history of close inter-action, they become problematic as ransomware groups seek to enlarge their footprint. The ambition to grow, combined with the insufficient vetting process and the ambiguity that shrouds the identities of new recruits, significantly magnifies the potential for OPSEC failures. This expansion not only stretches the operational capacity of ransomware groups but also introduces substantial security vulnerabilities, laying fertile ground for breaches in opera-tional security as the organization scales.

Moreover, ransomware groups heavily rely on communication and attacking infrastructure that is frequently not under their con-trol, often located in countries that differ from their own opera-tional base or are not affiliated with 'friendly' governments. This reliance introduces complications, and this dependency complicates their operations. For example, when a ransomware group sets up its command-and-control server in a foreign country, there is an increased risk of that server being intercepted or compromised. One case illustrating this risk is the dismantling of the Emotet bot-net, where the Dutch National Police's criminal investigation led to the discovery of a wealth of sensitive information, including com-promised email addresses, usernames, and passwords, all stored on Emotet's command-and-control server located in the Netherlands.[21] Moreover, ransomware groups often utilize communication plat-forms or conduct their financial dealings through cryptocurrency exchanges associated with entities in various countries, which can further jeopardize their operational security.[22]

4

DATA

"I am convinced that there are only two types of companies: those that have been hacked and those that will be. And even they are converging into one category: companies that have been hacked and will be hacked again". Robert S. Mueller III, former Director of the FBI and later also Special Counsel into the Russian interference of the USA election, said this in a keynote at the RSA Cyber Security Conference in 2012.[1] It is a variation of the often-heard quote, "There are only two types of businesses: Those that have been hacked and those who don't know they have been hacked". Given that Conti operates with the structure of a business entity, they, too, are not immune to the risks of data breaches.

In this chapter, to start my case study analysis, I explore the various data sources that have been instrumental in analyzing the Conti ransomware group. A significant portion of my insight comes from an extensive collection of internal messages and documents that were leaked following Russia's further invasion of Ukraine in February 2022. Alongside this pivotal leak, the study also draws upon a range of other critical sources that are essential for understanding Conti, as discussed in the preceding chapter. These sources encompass technical data extracted from threat intelligence reports, transactional details from cryptocurrency analysis platforms, findings from government publications, and information directly from Conti's own leak site.[2]

Danylo's Leaks and More

Danylo, a pseudonym, claims that he is a male Ukrainian computer specialist and student of the cybercriminal ecosystem. He recounts his initial penetration into the computer systems and networks of the group later identified as Conti back in 2016. In a conversation with *CNN*, Danylo shared that he had been covertly monitoring the hackers' servers for years, forwarding intelligence about their activities to European law enforcement. "Sometimes they make mistakes", he observed, highlighting the importance of seizing those moments. "I just was in the right place at the right time. I was monitoring them".

However, Danylo's perspective shifted dramatically on February 25, 2022, in response to a declaration posted by Conti on their website, Conti News. The message read:

```
"WARNING"

The Conti Team is officially announcing a full support
of Russian government. If any body will decide to
organize a cyberattack or any war activities against
Russia, we are going to use our all possible resour-
ces to strike back at the critical infrastructures of
an enemy.
```

"To prove that they are motherfuckers",[3] Danylo chose to publicly release the information he had collected on Conti by exploiting a vulnerability in the backend of Conti's XMPP chat server. To disseminate his findings, he established an X handle, @ContiLeaks, and began periodically posting links to the data hosted on Anonfiles, a platform allowing anonymous and free file uploads.[4] His X feed was not limited to these leaks; Danylo also expressed his condemnation of the Russian invasion and the resulting humanitarian crises, highlighting personal losses and the broader impact on civilians. In one of the posts Danylo states, "Fuck Russian inviders!" and "more sanctions! they destroy hospitals, and a lot of ppl died! even some of my friends!"

To alert the media and ensure widespread dissemination of his findings, Danylo reached out to journalists with a message that emphasized the significance of the leak and its authenticity:

DATA

```
Greetings,

Here is a friendly heads-up that the Conti gang has
just lost all their shit. Please know this is true.
https://twitter.com/ContiLeaks/status/1498030708736
073734

The link will take you to download an 1.tgz file that
can be unpacked running tar—xzvf 1.tgz command in
your terminal. The contents of the first dump contain
the chat communications (current, as of today and
going the past) of the Conti Ransomware gang. We
promise it is very interesting.

There are more dumps coming, stay tuned.
You can help the world by writing this as your top
story.

It is not malware or a joke.
This is being sent to many journalists and resear-
chers.

Thank you for your support.

Glory to Ukraine!
```

Danylo conveyed to *CNN* that a Russian airstrike had occurred perilously close to a relative's home. Having grown up in Ukraine during its time as part of the Soviet Union, Danylo was determined to prevent his country from reverting to Russian control.[5]

The initial batch of documents released by Danylo consisted of Jabber files, a communication platform formerly offered by Cisco and succeeded by WebEx. This platform supports messaging along with video and voice calls. The leaked files unveiled more than 60,000 internal messages spanning from January 21, 2021 to February 27, 2022. In subsequent days, Danylo continued to expose further information from Conti, including a file with 107,000 internal Jabber messages dating back to June 2020, coinciding with the initial deployment of the Conti malware.

The Conti ransomware group developed a centralized knowledge system with data and training material to ensure consistent work and avoid duplication. In early 2021, a Conti member named Alter advocated for the importance of such a system:

> "@all Friends, the amount of experience accumulated
> together exceeds the capabilities of normal and struc-
> turing storage, it was decided to raise a simple
> forum engine to publish relevant guides and materials
> there that we all get in the process of work.
>
> A big request to everyone! In their free time—write
> to me for registration, and after it—write down some
> article there, on the topic in the PM we will decide
> who can do what.
>
> The forum will not be used as some kind of chat room,
> more like storing and replenishing the knowledge
> base, it will be useful for everyone, I tried to
> divide the navigation by topic for the time being
> and put the approximate headings of the first articles
> where we will port the material.
>
> Be responsible, you can't even imagine how many ques-
> tions of the same type are asked to each other every
> day! We can seriously save time for ourselves and
> our colleagues!"

Subsequently, Danylo also published a variety of reference materials used by Conti, encompassing technical manuals and management instructions, a collection of a dozen GitHub repositories
containing the group's internal software, arrays of stolen credentials and certificates, screenshots showcasing their toolkit, and a
collection of educational documents and videos. Additionally, he
leaked another set of messages from RocketChat, a communication
platform similar to Slack or Discord, favored by Conti's technical
teams. Particularly captivating for many security researchers was a
password-protected archive within the leaks, which contained the
source code for Conti's ransomware encryptor, decryptor, and
builder tools. Although Danylo did not disclose the password,
another analyst swiftly cracked it, thereby exposing Conti's ransomware source code to the wider public.[6]

In a desperate attempt, Conti sought to retract their statement,
insisting that they were not endorsing any government:

"WARNING"

As a response to Western warmongering and American
threats to use cyber warfare against the citizens of

> Russian Federation, the Conti Team is officially
> announcing that we will use our full capacity to
> deliver retaliatory measures in case the Western war-
> mongers attempt to target critical infrastructure in
> Russia or any Russian-speaking region of the work-
> ing. We do not ally with any government and we con-
> demn the ongoing war. However, since the West is
> known to wage its wars primarily by targeting civil-
> ians, we will use our resources to strike back if the
> well being and safety of peaceful citizens will be
> at state due to American cyber aggression.

Despite the initial exposure, the leakage of information continued, with newer Jabber messages emerging after the initial data release. This revealed that various members of the Conti team, despite being aware of the security compromise, persisted in using the vulnerable servers, failing to eliminate the Ukrainian infiltrator.[7] These subsequent leaks offer a unique window into the group's immediate response to the breach. Evidence surfaced of Conti's administrators moving to new communication platforms and erasing previous communication in a bid to mitigate further damage.

Danylo's operation was actually not the first time that files were leaked from Conti. Previously, in early August 2021, an affiliate known as m1Geelka took to the Russian underground forum XSS to share Conti's operational manuals, detailing the group's tactics, techniques and procedures.[8] This individual also posted screenshots revealing the IP addresses of Conti's Cobalt Strike servers, critical for managing remote commands during their cyber operations.[9]

Danylo's leaks also did not mark the end of the breaches against Conti. On March 4, while many researchers were still analyzing the leaks from Danylo, a new X profile by the handle @trickleaks emerged, declaring, "We have evidence of the FSB's cooperation with members of the Trickbot criminal group (Wizard Spider, Maze, Conti, Diavol, Ruyk)". This profile then proceeded to leak the internal communications among Trickbot members—a criminal group closely linked to Conti. In the span of two months, around 250,000 messages from thirty-five potential members were uploaded.[10]

Analyzing the Leaked Documents

These leaks have collectively unveiled a treasure trove of insights, revealing the complex dynamics and operational frameworks of this vast criminal consortium.[11] As highlighted by an analyst, "One of the biggest enigmas surrounding ransomware has been the operational modalities of these criminal syndicates. The Conti leaks have afforded the cybersecurity community an unprecedented peek into the mechanics driving these illicit activities". This wealth of information has sparked significant interest within the cyber threat intelligence community, leading to in-depth analysis and sharing of findings derived from these leaks.

The leaks also greatly helped me to write this book; however, navigating through them presents its own set of challenges. First, the chat messages are in Russian and tend to be informal with slang, dubbed Феня (Fenya), and colloquial language that is not always immediately understandable. As Misha Glenny, a British journalist on the history of cybercrime, writes, "The first line of defence of criminal hackers in Russia or Ukraine is always the ever-changing local slang".[12]

Therefore, analyzing these leaked chat messages demands in-depth research and a comprehensive understanding of the context, making it challenging to derive accurate conclusions.[13] Much of the information necessitates interpretation beyond the explicit text. Even within Conti, operators occasionally struggle with unfamiliar terminology, reluctant to seek clarification from peers. An example from the leaks reveals an individual named "taker" seeking explanations from "zulas", a backend developer on terms like "lero" and "dero". We know from external forensic analysis that the terms refer to specific source code components of malware.[14]

While language translation programs and automated text analysis tools provide a starting point for examining the chats, they introduce their own issues. The loss of context and inaccuracies in translation can lead to misunderstandings.[15] For instance, the word "jabber" is often erroneously translated as "toad", and "dough" actually refers to "money", highlighting the limitations of relying on these translation tools for a nuanced understanding of the conversations.

Most messages over Jabber were sent between July 2020 and November 2020. They capture conversations between members across all functions and ranks of the organization, using anonymous handles like "skywalker", "darc", "dominik", "demon", and "dollar".[16] Brian Krebs, an investigative reporter on computer security, noticed that the chat logs have temporal gaps during the same periods when Conti faced interference from law enforcement or other organizations. As Krebs writes, the temporal gaps "roughly correspond to times when Conti's IT infrastructure was dismantled and/or infiltrated by security researchers, private companies, law enforcement, and national intelligence agencies. The holes in the chat logs also match up with periods of relative quiescence from the group, as it sought to re-establish its network of infected systems and dismiss its low-level staff as a security precaution".[17]

Post-intervention, as the timeline in Figure 1 shows, there is a visible decline in messaging frequency and activity levels among the group's members, indicating a shift towards caution or the adoption of different communication platforms. This change likely stems from members' greater wariness or changes in their operational security measures. Consequently, this evolution in communication habits has obscured visibility into the group's internal

Figure 1: Number of Jabber Messages between Conti Members between July 2020 and March 2022.

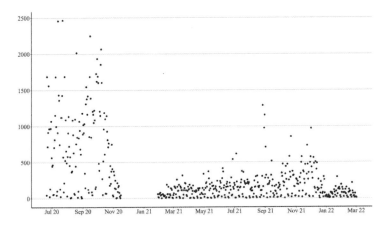

workings in later periods, potentially impacting the depth and accuracy of my analysis.

To grasp the nuances of the messages and gain insights into the ransomware group's operations, I tapped into a network of friends and colleagues for their expertise and perspectives. Additionally, I engaged with a wide array of professionals in the cybersecurity realm, including security researchers, ransomware negotiators, malware developers and other specialists, to deepen my understanding of the group's motives and behaviors. Throughout this process, I encountered a range of conflicting narratives, theories and evidence. In crafting this book, my goal was to shed light on these areas of agreement and disagreement, carefully assessing the strength of the evidence behind various assertions.

I also used a lot of other data sources, such as media reports with information about victim profiles and responses, posts on hacker fora, and Conti's own public writing and engagement. Sometimes the most detailed analysis is not found in a report of a billion-dollar cybersecurity company, but a personal blog of an aspiring researcher. For example, Chuong Dong, a Computer Science student at Georgia Tech, wrote a terrific analysis of how Conti exactly encrypts the files on a victim's machine on his personal blog.[18]

Leak Sites

In this book, I also explore Conti's operations through their leak site, notably Conti News. Similar to other ransomware groups, Conti uses this public website to publicize the data of victims who have chosen not to meet their ransom demands, an integral component of their double extortion method.

Leak site data is commonly used by practitioners and scholars to shed light on the geographic spread of victims and their organizational sizes, often without significant critical reflection on the implications or accuracy of this information. However, this use of this data necessitates a cautious approach.

A major limitation in analyzing leak site data is its selection bias, only showcasing victims who have not met ransom demands. For example, a highly successful ransomware group with a 60–70 percent payment success rate, due to their tactic of stealing vast amounts of critical data, puts immense pressure on victims to pay up. In contrast, another ransomware group might focus on minor third-party data in smaller quantities, leading to fewer successful ransom payments. Despite their lower success rate, this latter group may still be perceived as a leading ransomware entity, while the more effective group is ranked lower. This bias complicates cross-group comparisons, such as between Conti and its competitors.

This bias also challenges intra-group comparisons, making it difficult to discern trends of Conti and other groups over time. A decline in the volume of leaked data does not straightforwardly indicate a reduction in ransomware operations; it could also imply that more victims are yielding to ransom demand as the ransomware group became better in stealing critical data.

Additionally, ransomware groups have an incentive to manipulate this information for their advantage. Such manipulations could include timing data releases to inaccurately imply activity during specific periods, or inflating the volume of data to bolster their perceived threat level within the cybercriminal ecosystem. Conti, among others, has been observed to engage in both practices. The more the research community relies on data to observe trends, the greater the incentive for these groups to manipulate data.

Therefore, this study approaches leak site data with increased scrutiny, fully aware of its potential biases and the risk of manipulation. The goal is to dissect this data critically, understanding both its contributions and limitations in illuminating the operational dynamics and public interactions of ransomware groups.

Insider Access

The last source of data for my case study that deserves a deeper discussion is the use of insider access data. Insights gained from individuals with insider knowledge of Conti's operations have been helpful in understanding the group's internal dynamics, motivations, and the interplay between different factions within Conti and the broader criminal ecosystem. These insider perspectives have proven especially valuable when they spanned a considerable duration, providing context that enhances our comprehension of Conti's emergence. The emergence of leaks has also strengthened the validation of these insider accounts, which previously depended on the credibility of the sources or awaited confirmation from threat intelligence reports.

However, insider data comes with its limitations. Not all claims from insiders can be verified with the same level of confidence as other types of data. Furthermore, insider insights were more accessible and reliable in the earlier phases of Conti's operation when communication within the cybercriminal group was more centralized. Especially following Danylo's leaks, maintaining insider access became more challenging. The group's dispersion into smaller, more cautious factions, each operating with their own servers, made it difficult to obtain insights from the later stages of Conti's existence, presenting a significant constraint on our ability to capture the full scope of their operations.

5

THE ORIGINS

RYUK

In July 2019, hackers infected the computer systems of the city of New Bedford, a small city in Massachusetts with a population of about 100,000 people.[1] The attack took place on a Friday, just before regular working hours, a tactic often favored by cybercriminals. The hackers succeeded in infecting 158 workstations, approximately 4 percent of the city's computer network. The spread of the ransomware was limited thanks to the response of the city's staff. They swiftly contained the data encryption process by disconnecting servers and shutting down affected workstations, as praised by Mayor Jon Mitchell.[2]

The ransomware used against New Bedford was a variant of Ryuk. The attackers demanded a hefty payment of $5.3 million in Bitcoin for the data decryption keys' release. The city tried to negotiate for a more manageable amount of $400,000, but the offer was turned down, and the hackers did not present any new demands. The city then took matters into its own hands, attempting to recover the data without further negotiation.[3]

The attack on New Bedford was not the first appearance of Ryuk ransomware. As early as August 2018, reports appeared that a new "ransomware strain named Ryuk is making the rounds".[4] And then, within a relatively short period, from late 2018 to the second half of 2019, Ryuk swiftly claimed the title of the most used ransomware strain, surpassing even GandCrab as the most detected ransomware family.[5] The name Ryuk seems to be inspired by a character from a Japanese manga series called *Death Note*. In the manga, Ryuk is a Shinigami, a spirit overseeing the demise of humans.

Ryuk is now often cited as the forerunner of Conti. This chapter delves into Ryuk's origins to gain insight into the genesis of Conti. It highlights that ransomware groups do not spring from individuals with no experience in the business. Instead, they are often reformations of other groups, sharing some overlapping but not entirely similar membership. These entities might begin as small ventures, with ransomware operators eventually transitioning from one group to another to help set up the infrastructure and develop the necessary tooling. Operators, in turn, may rapidly shift their preference for a specific strain of ransomware. Understanding this evolutionary process sheds light on the complex landscape of ransomware operations and the interplay among these criminal entities.

Thank You North Korea

When Ryuk first appeared, various researchers attributed the ransomware to North Korea. This association was based on the observable code overlaps with Hermes ransomware, particularly evident in the encryption function for individual files and the scripts used for erasing backup files in both strains.[6]

Hermes, notably, had been adapted for use in an attack against the Far Eastern International Bank (FEIB) in Taiwan, an operation broadly attributed to the North Korean hacker collective known as the Lazarus Group. This group operates under the auspices of the Reconnaissance General Bureau of North Korea's military and has been implicated in numerous cyber offensives, including the WannaCry ransomware discussed previously.[7] In an operation spanning the end of September and the beginning of October 2017, the Lazarus Group launched a spear-phishing campaign to penetrate FEIB's digital defenses. Having infiltrated the network, they mapped out the infrastructure to locate computers with access to critical systems, subsequently deploying their customized malware. Their infiltration culminated in unauthorized access to the bank's SWIFT account, attempting to wire funds to banks across Sri Lanka, Cambodia, and the United States.[8] When FEIB discovered the fraudulent transactions, the Lazarus group installed the Hermes ransomware on the bank's network. This deployment was meant to delete evidence of their intrusion and impede investigations.[9]

Later developments led to a more persuasive theory regarding the relationship between Hermes and Ryuk, suggesting a common Russian origin for both ransomware families. This theory posits that the Lazarus Group from North Korea might have procured Hermes as a ready-made tool for their cyber operations, choosing to buy an existing solution instead of investing time and effort in creating a new one. An analyst of McAfee reflected on this possibility, stating, "What if the actor who attacked the Taiwanese bank simply bought a copy of Hermes and incorporated it into their campaign to create a diversion? Why bother building something from scratch when they can acquire the perfect distraction from an underground forum?"[10]

Supporting this theory, Hermes ransomware was listed for sale on the underground forum Exploit.in by a developer fluent in Russian in August 2017, showcasing a characteristic feature of Russian ransomware to exclude systems with Russian, Ukrainian, or Belarusian languages from encryption.[11] A later post in the same forum thread also mentions the ransomware Ryuk, possibly pointing to a connection between the two strains.[12]

While it is likely that Hermes and Ryuk originated from the same developer, their operational nuances underscore Ryuk's evolution. A key difference between them lies in Ryuk's systematic shutdown of specific applications and services prior to launching an attack on a system.[13] Furthermore, Ryuk adopts a more discerning strategy in its ransom demands, as evidenced by CheckPoint's analysis revealing varied ransom notes across different Ryuk attacks. Some victims were presented with elaborate, detailed demands, whereas others faced more succinct requests. Notably, the comprehensive demands typically associated with higher ransom fees (around 50 Bitcoin, or approximately $320,000) contrast with the less detailed ones asking for smaller amounts (15–35 Bitcoin, or about $224,000).[14] This variability in ransom notes indicates a tactical deployment of Ryuk, suggesting that it is selectively used post-network infiltration rather than disseminated via broad email spam campaigns.[15] In essence, Ryuk's development seems to reflect a more calculated, phased approach where insights from preceding phases inform the tactical adjustments in subsequent stages of the attack.

A Fatal Combo

In the earlier discussion on the history of ransomware, I highlighted Cryptolocker's pioneering use of the Gameover Zeus botnet to enhance its ransomware campaigns. Ryuk elevated this reliance on botnet infrastructure to a new level. From late 2019, it increasingly depended on two different software and infrastructures—Emotet and Trickbot—to create a fatal combination for its victims.

Emotet, initially surfacing in 2014 as a banking Trojan targeting banks in Germany and Austria, was designed to steal banking credentials from compromised machines.[16] This information was then exploited to facilitate unauthorized transactions to so-called money mule accounts, effectively laundering the illicit proceeds. Emotet distinguished itself through its propagation methods, primarily leveraging spam emails. It ingeniously harvested email contacts from infected users' Microsoft Outlook accounts, crafting convincing emails that were sent to these contacts with malicious attachments or links, thereby perpetuating the malware's spread. In some cases, Emotet also tried to crack passwords using brute force methods or vulnerabilities to gain further access to the system or install so that it could spread without direct user interaction.[17]

Over time, Emotet transcended its initial purpose, morphing into a more versatile "dropper". This term denotes its function to deploy and execute secondary malware payloads. By the time of Ryuk's adoption of Emotet, its payload module comprised a packed file with the core malware component and an anti-detection module. Upon infection, Emotet follows a set procedure: it relocates to a specific directory, generates a shortcut in the start-up folder to ensure persistence, collects and sends information about the infected machine to its command and control (C2) server, and, finally, is capable of retrieving and executing additional malicious payloads from the C2 server.[18]

TrickBot also started as a banking Trojan in 2016, and equally evolved into a modular malware with diverse capabilities.[19] The developers behind TrickBot are recognized for their innovation, continually enhancing the malware with new functionalities to steal credentials, expand network access, disable antivirus tools, or deploy other malware.[20] A key feature of TrickBot is that it is excellent in discovering and exploiting vulnerabilities within networks, making it extremely effective in setting the stage for follow-on ransomware attacks.[21]

Ryuk combines Emotet and TrickBot to enhance its impact. It typically begins with Emotet, which acts as an entry point, providing access to TrickBot. Once inside the network, TrickBot gathers key information and employs various capabilities for different tasks, facilitating lateral movement within the system.[22] In the last step, the Ryuk ransomware is released to encrypt the hard disks. The attackers also make sure to delete of any discovered data backups.[23]

It makes a lot of sense for Ryuk to rely on these powerful malwares to maximize its success. It does not have to worry about the entire attack chain and gains access to continuously updated tools. This partnership benefits both sides. The banking Trojans were not bringing in as much money anymore because banks got better at detecting fraudulent transactions. Also, running these operations requires a lot of technical upkeep, including updating web-injects and managing the botnet, as well as human involvement in handling money mules. Switching to ransomware makes it easier to make money. Instead of dealing with the hassle of moving money through banks and mules, attackers just get victims to pay them directly with cryptocurrency.[24]

Ryuk has raised the stakes in ransomware by demanding much higher ransoms over time, thereby capturing a larger slice of the market.[25] Starting in early 2020, it began targeting big businesses, pushing them towards hefty ransom payments to retrieve their data. Approximately twenty companies were targeted each week, with the typical ransomware payment reaching around $750,000 in Bitcoin.[26] In November 2020, it was reported that Ryuk's operators earned an astonishing $34 million from a single victim—a large company—in exchange for the decryption keys.[27]

Changing of the Guard

At the end of 2019, when Ryuk was still causing major havoc and was widely deployed, the Conti ransomware was observed for the very first time.[28] ID Ransomware, a platform allowing users to identify ransomware strains by submitting ransom notes or encrypted files, provides valuable data on this shift. Early in 2020, Ryuk remained prevalent, but by April there was a noticeable decline in Ryuk incidents, while submissions for Conti began to increase.[29]

Right from the outset, Conti was not shy about its connection to Ryuk, making no attempts to hide it. This was particularly evident in the ransom note used by Conti, which bore a striking resemblance to the template employed by Ryuk in previous attacks.[30]

Various terms have been used to describe the relationship between the two groups. The most common one is viewing Conti as the "successor" or "descendant" of Ryuk.[31] However, neither term perfectly captures the essence of their relationship, as both groups coexisted for a period and learned from each other. Moreover, some members were part of both groups, further blurring the lines. The most accurate way to portray the connection is that Conti emerged as a spin-off from Ryuk. As the spin-off gained momentum and success, more individuals transitioned from Ryuk to Conti, solidifying their bond.

The links between Ryuk and Conti appear to extend throughout all levels of their organization, encompassing operators who conduct operations for both, managers who oversee projects for both, and even high-level leaders who work for and receive commissions from both groups. In fact, Chainalysis has demonstrated that Stern, the leader of Conti, is known to have received payments from both Conti and Ryuk.[32] Additionally, Ian Gray and a team of researchers extracted the Bitcoin addresses mentioned in the Conti leaks, and then leveraged access to the Crystal Blockchain platform to analyze transactions from Conti operators. They revealed that Conti used leftover money from Ruyk to fund their operations, also showing that traceable income ($77.9 million) is more than double total expenses ($31.2 million).[33]

Similar to Ryuk, Trickbot is crucial in initiating infections and facilitating the deployment of Conti ransomware onto victims' systems. The technical and organizational ties between Trickbot and Conti are well-documented.[34] Analysis of leaked data reveals that about 18 percent of aliases used in Trickbot communications are also found in Conti's Jabber chats. Among these linked individuals is Alla, a middle-aged woman associated with Trickbot, who was detained by U.S. authorities in June 2021.[35] Known by the alias "Max", she is also the probable developer behind a ransomware variant discussed in chapter eight on Diavol, illustrating the intertwined relationships within these cybercriminal ecosystems.[36]

The integration of features from the Ryuk ransomware strains into Conti's operations was a calculated move to advance the group's capabilities. During a conversation captured in a Jabber chat on July 9, 2020, Stern, identified as Conti's leader, tasked Buza, the technical manager overseeing the coding team and product development, with finding samples of Ryuk "from the internet".[37] Stern instructed a developer to create an encryption tool inspired by the ransomware.

However, the exchange of tactics, techniques and procedures between Conti and Ryuk was mutual. Conti adapted methodologies from Ryuk, which, in turn, incorporated aspects of Conti's operations into its own. On October 8, 2020, the DFIR Report, a group of volunteer analysts specializing in investigating cyber intrusions, published a report titled "Ryuk's return", providing a comprehensive breakdown of the group's modus operandi.[38] A couple of days later, Buza shared the DFIR Report's report in their internal jabber chat with "Professor", another senior manager overseeing Conti's hacking operations. Professor's response was intriguing, stating, "well, not much different from our movements". Buza concurred, replying, "yes, practically nothing". Professor then noticed a specific batch file, "adf.bat", recognizing it as his own creation.

In this setup, where there is a partial membership overlap between Ryuk and Conti at various levels—from affiliates conducting attacks for both groups, to developers and even managers working for both—not everyone is fully aware of the relationship between the two. As a result, some individuals might be surprised to discover more code overlap than they had initially anticipated. To stay updated on each other's activities, the ransomware operators sometimes rely on publicly-disclosed investigations from the threat intelligence community. For example, Target mentioned to Stern that "Kremez told us that Ryuk infections have slowed down lately, as the threat actor is likely in a vacation kind of state". This information was obtained from AdvIntel's blog, a cybersecurity company with a reputation for possessing profound insights into the structure of these ransomware groups, and Vitali Kremez was co-leading that company at the time.[39]

Conti as the Leading Ransomware Group

In late 2019, Conti ransomware emerged while Ryuk was still active. Throughout 2020, as Ryuk submissions decreased, Conti's presence grew, establishing it as a spin-off that shared members and financial resources with Ryuk. Both groups exchanged tactics and personnel, facilitated by the use of common tools and infrastructure like Emotet and Trickbot for initiating attacks.[40]

It then did not take long for Conti to become the world's most profitable ransomware group. In 2021, over 400 organizations were successfully targeted by the ransomware group. Its estimated revenue was more than the combined earnings of its five main competitors: DarkSide, Phoenix, REvil, Cuba, and Clop.[41] The subsequent chapters will explore the mechanisms that enabled Conti to successfully establish itself in the ransomware landscape.

6

THE PLAYBOOK

In spring 2021, Conti used a phishing email with a compromised Excel file to infiltrate Ireland's Health Service Executive (HSE), which employs 130,000 people and is one of the country's largest employers.[1] Gaining initial access to an HSE computer, Conti spent a month undetected, exploring and spreading within the network, including HSE-affiliated hospitals and the Department of Health. Although an alert on May 12 warned of malicious activity, only the Department of Health and one hospital prevented the ransomware's execution. By the early hours of May 14, 2021, approximately 80 percent of HSE's data was encrypted.[2]

Reports of encrypted systems quickly reached the HSE's Office of the Chief Information Officer. Within hours, a Critical Incident Response Plan was enacted, and by early morning, HSE's servers and access to the National Healthcare Network were shut down, leaving all systems, even unaffected ones, inaccessible. Staff were advised to power down their computers, significantly disrupting services that relied on digital access, like radiology and lab work. The incident did not immediately threaten life as medical devices remained functional, but the lack of access to digital records caused significant operational disruptions, leading to service cancellations across 31 of 54 hospitals.[3]

Conti left a ransom note demanding $20 million to prevent the leak of 700 GB of sensitive data.[4] Despite this, the Irish

Prime Minister, Micheál Martin, declared that they would not pay any ransom.

Then, on May 20, Conti unexpectedly provided the decryption key via their negotiation chat portal but continued to threaten data leakage.[5] However, they never followed through with these threats, and no data was leaked even 18 months post-attack.[6] The lack of action may have been influenced by the widespread condemnation of the attack, including a statement from the Russian Embassy in Ireland offering assistance in the investigation.

Despite initial safety concerns, the decryption key was tested and applied by Mandiant, a cybersecurity company now part of Google Cloud, significantly aiding recovery efforts, though it still took nearly three months to fully restore systems.[7] The attack, amidst the COVID-19 pandemic, significantly stressed the already burdened healthcare staff. The attack's direct and indirect financial impacts were substantial, costing the HSE €51 million immediately, with projected cybersecurity improvements set to cost up to €657 million by 2029.[8]

In December 2021, the Irish Police's Cyber Crime Bureau succeeded in taking down several domains that were involved in the attack on the HSE, claiming that in doing so they were able to "directly prevent a large number of further ransomware attacks across the world".[9]

The story of Conti's attack on the HSE raises several overarching questions about the modus operandi of the group.[10] How does Conti choose its targets? How long does it typically get access and remain in a system? And how does it interact with its victims?[11]

This chapter describes the playbook that that propelled Conti to unprecedented success, dissecting the group's tactics, techniques and procedures into eleven distinct but interconnected stages, depicted in Figure 2.[12]

I explain that their process begins with reconnaissance, broadly scanning for exploitable openings rather than meticulously selecting targets based on vulnerability and value. While Conti employs a variety of techniques to compromise networks, phishing stands out as its preferred method, indicating a reliance on exploiting human vulnerabilities. Post-compromise, the group invests con-

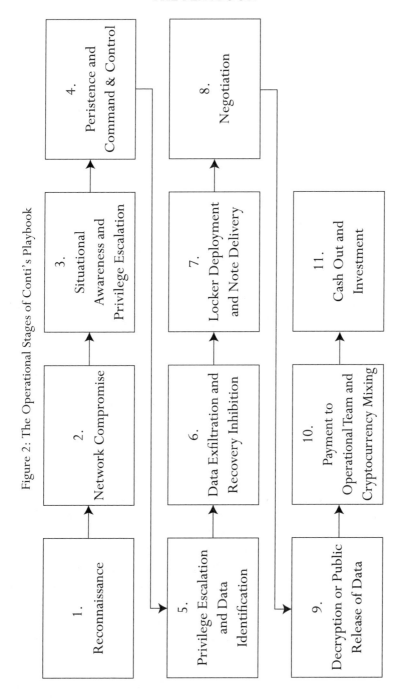

Figure 2: The Operational Stages of Conti's Playbook

siderable effort in mapping out the digital landscape of their targets, facilitating situational awareness and privilege escalation to secure comprehensive access and control. The group's identification and exfiltration of key data demonstrate a prioritization of high-value assets, ensuring that their operations impact is maximized. By stealing data, deploying encryption and subsequently dropping specific ransom notes, Conti lays the groundwork for complex negotiations, often leading to substantial ransom payments. The subsequent stages of payment processing seek to leverage the relative anonymity and flexibility of cryptocurrency. Their approach to cashing out and reinvesting speaks to a broader vision of sustaining and enhancing their criminal enterprise.

Writing this chapter was made a lot easier by the fact that, as I mentioned in chapter three, Conti has consolidated some of its best operational practices in one place. Conti curated the information in a channel called 'manuals_team_c' within the Rocket.Chat communications platform. Here, Conti outlined sixteen procedures used throughout their operations, spanning from the initial reconnaissance stage to the eventual exfiltration and encryption of data on the victims' machines. As part of the leaks, there were also two other relevant documents, translated as 'the hacker's quickstart guide' and 'the researcher's quickstart guide'.[13]

1. Reconnaissance

The initial phase of a Conti ransomware campaign involves reconnaissance, where the group gathers intelligence on prospective targets. This stage is crucial for identifying vulnerabilities and potential entry points into a target's network. Information collected during this phase can shed light on the target's organizational structure, key personnel, and network configurations. To facilitate this process, Conti has created an open-source intelligence (OSINT) team tasked with leveraging both open-source and commercially available tools to conduct their investigations.[14] The group acquired two accounts from ZoomInfo, a company that specializes in selling business data.[15] These accounts allow Conti to evaluate the size and revenue of potential targets, helping to calibrate ransom demands.

Despite these reconnaissance efforts, Conti's method for choosing victims does not seem as precise or deliberate as one might think. The OSINT team plays a more significant role in the post-compromise phase, helping to identify which entities have been successfully breached. There is little to suggest that Conti meticulously curates target databases. Rather, their approach to targeting is somewhat opportunistic, leaning more towards seizing opportunities as they arise, rather than following a pre-defined list of targets.

2. Network Compromise

The network compromise stage is the second step in Conti's playbook, focusing on infiltrating and establishing a foothold within the target's network. Conti leverages a variety of entry points for this purpose, as outlined in their internal "hacker's quickstart guide".[16] This guide serves as a comprehensive manual, suggesting various methods to breach networks, from exploiting public network services and vulnerable IoT devices to attacking web applications and dormant vulnerabilities in connected devices. IoT devices—like smart printers—are especially singled out for their infrequent updates, rendering them softer targets.[17]

The "hacker's quickstart guide" emphasizes gaining access to critical network components such as Active Directory and Domain Controllers. Achieving this access facilitates lateral movement across the network, allowing Conti to assert control over an increasing number of machines. The guide suggests using well-known pentesting frameworks like Metasploit, Cobalt Strike, Burp Suite, Puppy, and others, while also noting the effectiveness of custom tools developed by Conti to evade antivirus detection.

State-sponsored groups commonly acquire antivirus and security detection tools to evaluate the detectability of their capabilities and adjust their methods to evade defense mechanisms effectively. For example, *Recorded Future*'s Insikt Group found six procurement documents from PLA military websites and other sources, revealing that Unit 61419 of PLA's Strategic Support Force sought to purchase antivirus software from prominent American, European, and Russian security companies.[18] Conti also adopts this practice. Internally, Conti members recognized VMWare Carbon Black as the best endpoint security solution that could thwart their attacks. To tackle this, Conti tried multiple times to acquire a license for this software to aid their testing. In April 2021, they managed to get access through a proxy.[19]

Conti relies heavily on social engineering tactics to carry out their attacks.[20] The leaked information provides us with valuable insights into how these campaigns are orchestrated.[21] Two members of Conti, going by the aliases "Lemur" and "Terry", developed various phishing templates designed to target different organizations.[22] Moreover, (groups affiliated with) Conti later also reintroduced "BazarCall campaigns". Unlike traditional phishing methods that use malicious links or attachments, BazarCall campaigns encourage recipients to call a provided phone number. Upon calling, individuals are directed by operators who guide them through the process of unwittingly installing malware on their computer systems.[23] As Microsoft notes, this approach, which mirrors tech support scams, inverses the typical dynamic by having the potential victims initiate contact.[24]

Conti does not exclusively conduct these activities in-house. Similar to other ransomware groups, Conti relies on initial access brokers (IAB) for targeting and gaining access. The Google Threat Analysis Group (TAG) has reported on an Initial Access Broker named EXOTIC LILY, associated with Conti.[25] EXOTIC LILY predominantly employs spear phishing as its method for securing access, with Google TAG reporting that at its peak of operation activity the group was sending over 5,000 emails daily, targeting up to 650 organizations worldwide. Despite this broad outreach, EXOTIC LILY attacks appear to be sent by real human operators with limited automation. They were often targeted, employing tactics like employee spoofing to forge trust with the specific organization. In addition to spear phishing, EXOTIC LILY has been known to exploit a zero-day vulnerability in a browser engine for Microsoft Internet Explorer.[26]

3. Situational Awareness and Privilege Escalation

In the third stage of the attack process, Conti focuses on situational awareness and the elevation of privileges within the compromised environment. The goal is to find vulnerabilities that can be exploited to grant the attacks elevated access, thereby consolidating their hold over the system or network. Researchers from Tenable, diving into the vulnerabilities detailed in the Conti leaks, found that approximately 75 percent of these vulnerabilities target privilege escalation rather than initial access. This makes sense since Conti can find different ways to enter an organization aside from vulnerabilities, but they have fewer options to escalate privileges.[27]

Conti and its affiliates deploy a range of techniques and tools for lateral movement across the network, with Cobalt Strike being a particularly favored tool.[28] Launched in 2012, Cobalt Strike is widely used by cybersecurity professionals, including red teams and penetration testers. Its popularity extends to ransomware groups and other malicious actors as well. Cobalt Strike is essentially a versatile toolkit that allows attackers to conduct simulated network attacks and maintain stealth via hidden communication channels.

4. Persistence and Command and Control

Once Conti hackers escalate their access within a system, their top priority is to maintain control and avoid detections. To this end, they employ a variety of tools, such backdoors, webshells, and remote access software, which grant them comprehensive control over the compromised system and its resources, all while mimicking normal operations to elude detection.[29]

In addition to leveraging readily available commercial tools, Conti's technical documentation highlights the use of bespoke tools tailored to their specific needs. Among these, the Anchor backdoor stands out for its sophistication. Believed to have been developed by the Trickbot team, Anchor is particularly elusive due to its use of the DNS protocol for its command-and-control communications, making it a great tool for maintaining stealthy, long-term access to infected networks.[30]

5. Identification Directories with Relevant Data

At this stage, Conti focuses on identifying crucial directories containing important data. This normally begins with prioritizing administrators due to their access levels to critical systems and information. A section in the trickconti forum, "Hunt Administrator", outlines strategies for identifying and prioritizing network users with administrative privileges.[31] Hackers run various queries and then (manually) look for clues such as group memberships, department affiliations, or job titles. They verify the accuracy of their search results by examining the account statuses and last login time, and sometimes cross-referencing with LinkedIn profiles.

Next, they determine where these high-value accounts are active within the network, often utilizing the PowerSploit tool's Find-DomainUserLocation function. This enables them to impersonate these users, using impersonation tokens to remotely access and extract files.[32]

The final step involves searching through commonly used user directories, such as OneDrive or Documents, for sensitive files, including password lists. They also explore browser data, such as Firefox or Chrome browsing history and login information, to uncover further valuable information, like the locations of backup and virtualization servers.

6. Exfiltration and Analysis of Data

Following their identification of valuable data, Conti proceeds to exfiltrate it from the compromised organization's network. They use compression and encryption to package the data discreetly, minimizing the risk of detection during the transfer process. One of the tools Conti adapts for this purpose is Rclone, initially intended to manage backups to cloud storage. The group repurposes Rclone to move data stealthily to cloud services such as MEGA. Detailed guidance on configuring Rclone for data exfiltration is available on the trickconti forum, explaining how to use the tool effectively to transfer data from compromised systems.[33]

Additionally, Conti employs FileZilla, a widely-used, free File Transfer Protocol (FTP) software. FileZilla facilitates the creation of FTP servers or connections to existing ones, simplifying the process of sending out exfiltrated data.[34] Its intuitive interface proves useful for managing file transfers, further streamlining the exfiltration process for Conti's operators.

7. Encryption and Inhibit Recovery, and Ransom Note

In the seventh phase of their ransomware operation, Conti shifts its focus to the pivotal action: encrypting the victim's data and hindering their recovery process. Typically, it takes around four days for Conti to get to this stage from the moment they have initial access to the victim's computer system.[35] The overarching objective here is to induce a state of urgency and panic in the victim.

Among the most significant and damaging pieces of leaked information regarding Conti is the code for their ransomware.[36] It has two key components: a locker for encrypting files and a decryptor. Conti uses advanced encryption algorithms like chacha20, with dynamically generated keys for each encryption instance, enhancing the security and variability of the encryption process.[37] The locker, notably the more complex of the two components, incorporates several security measures, including anti-debugging features.

Chuong, a Computer Science student at Georgia Tech, conducted a thorough analysis of Conti's encryption method, particularly focusing on the second version of the locker, as early as December 2020. As part of a side project alongside his academic studies, Chuong explored the nuances of Conti's file encryption approach and the sophistication of the code. He observed:

> the encryption scheme is fairly straightforward, involving a randomly generated key safeguarded by a publicly known RSA key. Nonetheless, the implementation of multi-threaded encryption is notably elegant, utilizing a shared structure among threads to achieve remarkably swift encryption speeds. It's apparent that Conti's creators prioritize speed over encryption quality, as they cleverly bypass encrypting large files. Furthermore, the ransomware actively seeks network shares for spreading through its networking capabilities.[38]

Following encryption, files are appended with a ".CONTI" extension. For instance, a file initially named "Max_secret_dutch_pancakes_recipe.docx" would be displayed as "Max_secret_dutch_pancakes_recipe.docx.CONTI" post-encryption. Once encryption concludes, a ransom note titled "CONTI_README.txt" is placed on the victim's desktop, detailing the demands.

Subsequent to the release of Conti version 2, version 3 was introduced. However, analyses suggest it might not represent an improvement.[39] Payload, a Polish magazine specializing in offensive IT security, considers Version 3 as a significant regression in terms of the quality of its code, and believes that it was developed by a less experienced coder.[40]

Once the encryption process is complete, Conti shifts its focus to compromising the victim's backup systems. By targeting backups for encryption or destruction, Conti significantly diminishes the victim's capacity to recover their data, consolidating the group's leverage.[41]

The ransom note has evolved significantly over time, becoming increasingly detailed. Initially, Conti opted for a straightforward message, similar to those used by Ryuk, cautioning against unauthorized decryption attempts and directing victims to contact them via provided email addresses:

```
Your system is LOCKED. Write us on the emails:
[…]@protonmail.com
[…]@protonmail.com
DO NOT TRY to decrypt files using other software.
```

This approach changed in August 2020 with the introduction of a leak site threat, marking Conti's foray into the 'double extortion' tactic, a method pioneered by The Maze Crew only a few months prior.

```
"Your system is locked down. Do not try to decrypt,
otherwise you will damage fails. For decryption tool
write on the email:
[…]@protonmail.com
[…]@protonmail.com
If you do not pay will publish all private data on […]
```

As Conti's operations progressed, their ransom notes became more elaborate. A commonly seen note outlines the encryption of files by the Conti strain and warns against the use of third-party recovery software due to the potential for further data damage. It offers the decryption of two files free of charge as proof of their capability, along with contact instructions through a dedicated website accessible via the Tor browser or a regular web browser. This note also includes a threat of data publication on Conti's news website if the victim chooses to disregard their demands:

```
All of your files are currently encrypted by CONTI
strain.

As you know (if you don't—just "google it), all of
the data that has been encrypted by our software
cannot be recovered by any means without contacting
our team directly.

If you try to use any additional recovery software—
the files might be damaged, so if you are willing to
try—try it on the data of the lowest value.

To make sure that we REALLY CAN get your data back—
we offer you to decrypt 2 random files completely free
of charge.

You can contact our team directly for further
instructions through our website:

TOR VERSION:

you should download and install TOR browser first
hxxps://torproject.org)

hxxp://[…].onion

hxxpS VERSION:

hxxps://contirecovery.best

YOU SHOULD BE AWARE!

Just in case, if you try to ignore us. We've down-
loaded a pack of your internal data and are ready to
publish it on out news website if you do not respond.
So it will be better for both sides if you contact
us as soon as possible.

---BEGIN ID---

[…]

---END ID---
```

8. Negotiations Between Conti and Victim

During the negotiation phase, the Conti team navigates a fine line to maximize profits while assessing the victim's capacity and willingness to pay. A key aspect of this negotiation is assessing whether the victim prioritizes decrypting their data or preventing its leak. A revealing exchange between Pumba, a Conti operator, and Tramp, his supervisor, in January 2022, illustrates this balance. Tramp initially proposes a bold $20 million ransom. However, Pumba interjects, positing that the victim's primary concern might be avoiding data leaks rather than decrypting data. This shifts their focus toward leveraging the threat of data exposure. Tramp is surprised that this particular victim has not reached out and notes that the breach has only affected a few servers and systems. This observation leads Tramp to estimate a possible ransom range of $4 to $5 million. Pumba theorizes that the victim may have managed to restore their systems, possibly explaining their lack of communication. Advocating for a strategic approach, Tramp advises patience and suggests waiting for the victim's counteroffer.

The negotiation dynamics between Conti and their victims mirror a poker game where both sides play with hidden cards, each trying to gauge the other's true position. Despite efforts by Conti's OSINT team to ascertain a company's financial situation, much often remains uncertain. For example, a dialogue between Pumba and Tramp reveals Pumba's doubts about a victim's financial limits, noting, "I'm still in a fog, I can't really figure out if they're bluffing or if they really just stopped at 300k and that's it, their ceiling". Conti, too, employs bluffing tactics, threatening to leak data they may not actually have to pressure companies, much like a poker player feigning confidence with a poor hand.

At times, Conti's negotiations are bolstered by their relationships with specialized negotiators from firms dedicated to assisting victims of ransomware through the payment process. Among these negotiators, "the Spaniard", a Romanian individual working for a renowned ransomware recovery firm in Canada, proves to be an invaluable asset. On December 12, 2021, Mango, a senior Conti manager and the right hand of Stern, shares intel with his colleague

Bio about the efficient negotiation process facilitated by The Spaniard during discussions with LeMans Corp., a sports equipment company from Wisconsin. The Spaniard's expertise helps Conti streamline negotiations, with one communication highlighting their practical approach:

```
"They are willing to pay $1KK [$1 million] quickly.
Need decryptors. The board is willing to go to a
maximum of $1KK, which is what I provided to you.
Hopefully, they will understand. The company revenue
is under $100KK [$100 million]. This is not a large
organization. Let me know what you can do. But if
you have information about their cyber insurance and
maybe they have a lot of money in their account, I
need a bank payout, then I can bargain. I'll be
online by 21-00 Moscow time. For now, take a look
at the documents and see if there is insurance and
bank statements".
```

While The Spaniard often aids in securing favorable outcomes for Conti, their negotiations do not invariably meet Conti's initial demands. In one instance on October 7, 2021, The Spaniard communicated a firm limit from a client, stating, "My client can only afford a maximum of $200,000 and is solely interested in retrieving the data. If you cannot accommodate this request, the deal will not proceed".[42] An interesting tidbit from the internal communications reveals The Spaniard's travel plans, with a message dated August 16, 2021, from Mango to Stern mentioning "Spanishka's" upcoming visit to St. Petersburg to meet with friends. Mango amusingly considers befriending The Spaniard.

Conti also seeks to leverage relationships with other external parties, such as journalists. An internal discussion reveals a Conti member discussing the potential enlistment of a journalist to pressure victims into paying ransoms, with a proposed fee of 5 percent of the payout.[43]

Ultimately, the prevalence of cyber insurance among organizations as a measure to mitigate the financial impact of ransomware attacks presents a nuanced aspect of Conti's negotiations. The chat logs reveal that Conti has mixed feelings about targeting and negotiating with such victims.[44] While these policies can restrict the maximum ransom Conti might expect, they also mean that insured companies are generally more prepared to pay, often leading to quicker resolutions without prolonged negotiation.

9. Payment or Publication

The ninth stage of a Conti ransomware attack revolves around the victim's decision to pay the ransom for data decryption or face the consequence of having their data publicly exposed on Conti's leak site. Conti's ransom notes in later versions included a link directing victims to a webpage featuring a countdown timer. The use of countdown timers to pressure victims is not new and can be traced back to the Jigsaw ransomware in 2016, which intensified file deletion as the ransom remained unpaid over time. If the $150 ransom was not paid within the first hour, the ransomware would erase a single file. As time passed, the deletion rate escalated, with more files wiped out each hour, and this tally would rise whenever the 60-minute timer restarted. Every time the program was rebooted, up to 1,000 files were being deleted.[45]

Should the victim decide to pay, Conti provides a decryptor tool, allowing for the recovery of encrypted data, and refrains from publishing any leaked data. Conti has an incentive to provide a functional decryptor as it supports their reputation of rewarding compliance, thereby encouraging future victims to pay the ransom.[46] To facilitate this, Conti has established a support team dedicated to assisting victims through the decryption process. The highest recorded ransom paid to Conti amounted to 725 BTC, approximately $8 million at the time, by CareerBuilder, a Chicago-based job search platform.[47] According to data from Coveware, a blockchain analysis firm, the average ransom payment made to Conti in the latter part of 2021 and early 2022 was around $750,000.

10. Money Transfers and Cryptocurrency Mixing

The next step for Conti involves distributing the ransom payments among its members and obscuring the trail of these financial transactions. Cryptocurrency, particularly Bitcoin, is the preferred mode of payment due to its global acceptance and relative ease of use. However, while the transparency of blockchain technology facilitates unstoppable transactions, it also allows for the potential tracing of these funds by law enforcement and analysts. For example, the tracking of Bitcoin transactions played a key role in the recovery of a significant portion of the ransom paid during the Colonial Pipeline cyberattack by the DarkSide group.[48]

There are available alternatives like Monero that provide greater transaction anonymity.[49] Whereas Bitcoin operates as a fully transparent system, allowing individuals to view precise transaction amounts as funds move between users, Monero conceals this information to safeguard user privacy throughout all transactions.[50] It achieves this through several key features. Ring signatures mix a user's transaction with others, making it challenging to pinpoint the actual sender. Ring Confidential Transactions (RingCT) keep the transaction amount hidden, and stealth addresses ensure that each transaction generates a unique, untraceable address. Despite its advantages for preserving anonymity, Monero has not become the primary choice for Conti.[51] The preference for Bitcoin persists due to its greater liquidity and ease of use among victims.[52]

Conti utilizes cryptocurrency mixing services, known as tumblers, to obfuscate the trail of their cryptocurrency transactions. These mixers amalgamate various crypto assets before redistributing them, complicating efforts to trace the assets back to their origins. Think of this process like a group of people swapping different colored marbles in a hat. Each person puts their marbles into the hat, and after a while, the marbles are shuffled around and redistributed randomly to the participants. This way, it becomes challenging to determine which person originally contributed which marbles, adding a layer of obscurity to the origins of the marbles. For their services, mixers charge a fee, typically between 1 and 3 percent of the total amount mixed, as their profit margin.

Conti's use of cryptocurrency mixing strategies does not always yield the intended results. Sometimes, their efforts to obscure the origins of their funds actually lead to a greater concentration of illicit funds. In some cases, their attempt to blend funds ends up

Table 3: Conti's Tactics, Techniques and Procedures

Tactics	Techniques	Procedures
1. Reconnaissance	• Open-source intelligence, network scanning, domain enumeration.	• Using tools like ZoomInfo to assess target size and revenue, and querying for domain-specific data.
2. Network Compromise	• Exploiting entry points (phishing, vulnerable services, compromised credentials).	• Employing tools like Metasploit Framework, Core Impact, Cobalt Strike, exploiting IoT devices, targeting Active Directory and Domain Controllers.
3. Situational Awareness and Privilege Escalation	• Identifying vulnerabilities, escalating privileges, accessing sensitive systems.	• Using tools like Cobalt Strike for lateral movement and hidden communication channels.
4. Persistence and Command and Control	• Using backdoors, webshells, remote access software.	• Implementing tools like the Anchor backdoor, using DNS protocol for communication.
5. Identification of Directories with Relevant Data	• Prioritizing administrators, imper-sonating users.	• Manual investigation, cross-referencing, data reconnaissance, browsing histories.

6. Exfiltration of Data	• Using compression and encryption.	• Packaging data, using Rclone for cloud storage, exploiting FileZilla's FTP capabilities.
7. Encryption and Inhibit Recovery, and Ransom Note	• Using locker.	• Employing encryption algorithms, locking files with ".CONTI" extension, deploying ransom notes.
8. Negotiations between Conti and Victim	• Engaging with victims for ransom payments, leveraging connections, bluffing.	• Negotiating ransom amount, involving external parties like The Spaniard, using journalist pressure tactics.
9. Payment or Publication	• Pressure tactics, data publication.	• Countdown timer pressure, offering decryptor, potential publication on leak site.
10. Money Transfers and Cryptocurrency Mixing	• Receiving ransom payments in cryptocurrency.	• Using cryptocurrency exchanges, employing cryptocurrency mixers to obscure transactions.
11. Cash Out and Investment	• Cashing out earnings, investing funds back into infrastructure.	• Using OTC platforms for fiat conversion, covering expenses, making strategic investments.

creating a scenario where they possess even more red funds than before. In other words, attempts to mix marbles of different colors to make their origins less traceable can lead to a larger collection of the stolen marbles, which paradoxically attracts more attention.[53]

Conti operatives pay themselves out through cryptocurrency exchanges. By analyzing conversations leaked from the group, researchers were able to track the movement of ransom payments associated with both Conti and Ryuk.[54] Interestingly, a significant number of these transactions were made through exchanges known for their compliance with Know Your Customer (KYC) regulations, like Gemini and Binance. These platforms offer security for handling large transactions. However, this approach could inadvertently highlight the massive amounts of money being processed, drawing attention to Conti's activities and potentially compromising their operational security. Other funds are transferred to less reputable platforms, including the Seychelles-based Huobi exchange and the sanctioned Hydra marketplace, indicating a diversification in their money laundering tactics.[55]

11. Cash out and Investment

The final stage encompasses the delicate tasks of cashing out the illicit earnings and making strategic investments. New members receive instructions, typically from Stern or another senior figure, on how to convert Bitcoin into fiat currency. They are directed to use the Russian website Bestchange.com to identify an over-the-counter (OTC) exchange offering good rates. Recruits must then place an order to sell Bitcoin, opting for a transfer to a debit card as the payment method. After generating a deposit address, they share it with Stern, who completes the transaction, allowing the recruit to receive money directly on their debit card.[56]

Additionally, a portion of Conti's income is reinvested into the group's operational costs. Conti also allocates a part of its funds for covering operational expenses such as cloud services and software licenses.[57] For example, a technical lead named "Defender" asks Stern for $700 in Bitcoin to finance the group's server costs. These expenses pose a challenge because IT products often require payment in conventional currency. In a dialogue, system administrator Strix queries Carter on how he manages to pay for server services via PayPal, noting the difficulty of using over-the-counter brokers for small amounts like Strix's seven euros monthly server fee.[58] Carter explains his method of transferring cryptocurrency from LocalBitcoins to a "phantom" debit card—not registered under his real name—to facilitate anonymous, secure payments.[59] This debit card is linked to a verified PayPal account.[60]

The TTP Blueprint

Conti's tactics, techniques, and procedures (TTPs), as outlined in Table 3, indicate a comprehensive blueprint that spans from reconnaissance to cash-outs. This blueprint underscores the group's adaptability and learning curve, incorporating structured data gathering, exploiting vulnerabilities for network compromise, and employing social engineering to facilitate their operations. The negotiation process and interaction with victims through crafted ransom notes and payment demands highlight Conti's nuanced understanding of gaining leverage.

7

ORGANIZATIONAL STRUCTURE

The release of Danylo's leaks marked a rare opportunity to dissect the inner workings of Conti, an analysis typically impossible with conventional sources like malware samples and cryptocurrency transactions. As many pointed out, it revealed that Conti exhibits clear hierarchies and operates with a structure that resembles that of startups. At the same time, Conti's operations were deeply inefficient: tasks were often mismanaged, communication was fraught with breakdowns, and job satisfaction among its members was notably low. These organizational inefficiencies have predominantly been attributed to a lack of managerial expertise within Conti. Journalist Brian Krebs, for example, highlights the general dearth of managerial expertise among Conti operators.[1] This sentiment is echoed by Allan Liska, who notes: "They're not like senior executives or seasoned operators or things like that. These are people in their 20s and 30s that are running them and clearly have no concept of how to manage a large organization like this. Everyone [thinks] it's easy to be a manager. It really isn't".[2]

In this chapter, we delve into the organizational framework of Conti, breaking it down into four fundamental aspects. I begin with an examination of Conti's hierarchy, tracing the distribution of roles from senior leadership to those on the ground. I discuss the central role of Stern, who functions as the de facto CEO, in steering Conti's criminal activities and operational direction. I also address the division of labor among specialized roles such as malware developers, hackers, and negotiators, and how this hierarchy underpins the group's coordinated and illicit operations. My attention then turns to the group's workspace and workflow patterns. Here, similarities with conventional workplace expectations emerge, including regular updates, defined work hours, and discussions on remote versus office-based work arrangements. These insights shed light on Conti's attempts to maintain a semblance of normalcy within its clandestine operations. The narrative then progresses to recruitment and onboarding in Conti, examining the strategies employed to attract, vet, and integrate new members into the fold. I discuss the organization's ongoing battle with a high rate of attrition. This constant turnover necessitates a continuous effort to attract and integrate new members, underscoring the importance of efficient and effective recruitment strategies. Finally, the chapter examines Conti's compensation and reward system, showing how the group seeks to motivate its members through a performance-based reward system. However, this system is applied inconsistently by Stern and his senior management.

Organizational Hierarchy

In this section, I delve deeper into the hierarchical structure of Conti, concentrating on the role distribution from the top leadership down to the operational staff. At the heart of Conti's organizational structure is a figure known as 'Stern'. Journalist Brian Krebs suggests that Stern's name is quite apt, given that he "incessantly needled Conti underlings to complete their assigned tasks".[3] Stern's role goes beyond just leading Conti; he also holds key positions in affiliated groups like Trickbot.[4] Stern's exact relationship with the Ryuk ransomware team remains a topic of debate.

Stern's authority within Conti is unmistakable, largely because he controls the group's finances. He is the one who sends out salaries and approves budgets, effectively holding the purse strings. This financial control is a classic hallmark of leadership, giving Stern the power to steer Conti's direction and ensure that its operations run smoothly. As we will see later, this financial control also introduces a significant vulnerability to the group.

As we dive deeper into the organizational structure of Conti, Stern's role as a central figure becomes even more apparent. His leadership and financial oversight are critical for maintaining the group's operational efficiency, driving its expansion, and navigating relationships with other entities.

Towards the end of 2020, Conti was still in what could be described as its startup phase. Together with a small group of senior managers, Stern was at the forefront of this period, putting considerable effort into assembling the right team, honing efficient processes, and securing the technology essential for their ransomware endeavors. The content of leaked messages from August 2020 gives us a peek into these foundational efforts.

In his messages, Target shares with Stern the difficult yet fruitful challenges they are encountering, noting, "So far, things are going very intense, very difficult but productive". Target is pleased with "the interim phase of building our foundation". He further outlines to Stern how investments are being channeled:

```
right now the money is flowing in three directions /
n1) it's operators running costs + expansion = total
2 officers with large teams—one core and one new one
in training /n2) hackers offices (3 pcs)—interviews,
tech, rent, interviews, deposits, servers inside,
equipment, hiring and help hiring and shit else, and
after a week more payroll will be added for those
who will work there (20+ hackers)/n3) office with
programmers and equipment for them + team leader
already hired good and he will build the team for
the prof, this is an important devops for the prof,
the prof is happy with everything and he needs it
very much /n+specialists hired by the prof, to speed
up various processes /nhy sure it will pay off, so do
not get nervous.
```

Target also conveys to Stern a sense of optimism about the outcomes of their current efforts, suggesting that despite the high stakes and the pressures involved, they believe their focused approach to building a skilled team and robust operations will distinguish Conti in the competitive ransomware market. Target later also writes to Stern, "if we don't go crazy with everything, the prof convinced me that no one will be cooler than us neither in scale of work nor in results".

A discussion captured in Rocketchat reveals insights into Conti's operational setup as of late August 2020, highlighting an already established division into three main teams within the operations department.[5] Alter writes:

```
The structure is the following.
The current composition is divided into groups, each
group is assigned a team leader (one or two, depend-
ing on the size of the group).
Ateam—team leader rozetka
Bteam—team leader red and ali
Cteam—team leader steven
Team leaders are responsible for:
1. Issue cases for work
2. Teach, advise, instruct
3. Connect in the process of solving atypical or
   previously not completed tasks
4. Help with load builds, networking and other tech-
   nical issues related to software
5. Provide the necessary guides and manuals
The working group is required to:
1. Listen
2. Watch
3. Do
4. Learn
5. Ask questions
6. Follow the guides and instructions, complete the
   assigned tasks.
```

Target's optimism in late 2020 regarding Conti's competitive edge in the market appears to have been well-founded. By July 2021, Conti had expanded its workforce to roughly a hundred individuals. Mango provides Stern with a detailed breakdown: the core team comprises sixty-two members (fifty-four of whom are salaried), a reverse engineering team of twenty-three, a new coding team of six (four salaried), six reversers, and an Open-Source Intelligence (OSINT) team of four. Within this core team are operators who execute the operations. They also often engage in the initial communication with the victims and lead the ransom negotiations. Part of this core team also includes staff dedicated to marketing and communications roles, responsible for managing the leak site and posting on hacker forums. The group's more technical personnel, although smaller in number, play a key role in developing, refining, and maintaining the malware, including ransomware lockers and decryptors, essential for carrying out Conti's operations. The OSINT team is focused on researching potential targets, identifying vulnerabilities, and identifying possible access points.

Mango explains to Stern that the monthly operational costs, including expenditures for servers, security, and onboarding tasks for newcomers, totals around $164,800. Conti's staff must demonstrate their skills and commitment before being eligible to receive a profit percentage from their operations. This system, reminiscent of the partnership models in law firms, aims to ensure that significant contributors are rewarded with earnings beyond a basic salary. However, the specifics of the payment and incentive scheme can sometimes remain opaque to the employees. This ambiguity is highlighted in an exchange where "Bio" informs "Skippy": "as I understand and there will be a salary, and then as the payments go they will transfer to a percentage. [...] and how much salary here I don't even know, they didn't tell me at all".

Unlike many ransomware entities that lean towards a RaaS model, Conti has taken a different route by more emphasizing an in-house operational framework, supported by a large team of full-time employees. This approach underscores Conti's preference for maintaining tighter control over its operations. Initially, there was contemplation within Conti about pivoting to a more RaaS-focused model. However, this idea was eventually shelved, likely in response to the operational risks highlighted by the leaks from m1Geelka in August 2021, one of Conti's select few affiliates.[6]

While Conti does not heavily depend on the RaaS model for spreading its ransomware, it does outsource various activities when it sees fit. One example is their use of initial access brokers to gain entry into target networks. I previously discussed Google's TAG publication on Exotic Lilly an Initial Access Broker working closely together with Conti, for providing such access.[7] Conti also uses money mules for some of its money laundering efforts to uphold operational security.

Workspace and Workflow

Most of Conti's team operates remotely through 'home office' arrangements, but Conti does maintain physical offices. This setup has sparked debates within Conti that mirror mirror other companies' (post-)COVID discussions about the effectiveness of remote versus traditional office work. Target leans towards the benefits of an office setting, arguing it allows for better managerial oversight and tasks are done more efficiently. Yet Stern champions the effectiveness of remote work, challenging the assumption that physical presence in an office boosts productivity. He writes to Target, "[you say that] what coders online take weeks to do, coders offline do in days. It's not like that, I've had coders offline. It's the same in actuality". Those working remotely, according to Stern, "just need to be monitored every day" by managers.

Interestingly, these internal debates overlook two crucial aspects of choosing between remote and office-based work setups. Firstly, there is a remarkable lack of consideration about how their preferred working method could influence their security and vulnerability to surveillance. Secondly, they do not discuss how their work style might impact their ability to attract new talent; it is not clear whether individuals are interested in working in an office and whether this creates a challenge to recruit new hackers into the group.

Following Stern's managerial advice about remote work, team leaders expect their crew to provide regular updates. Everyone is supposed to be at their desk or online during certain hours. As one manager puts it, "work on time, be adequate and be in touch" and "working day from 9 to 18 Moscow time, you can +- start-end". This routine even shows up in when people are most active on Jabber, with most messages being sent around 2pm Moscow time.

Conti employees avoid working on the weekend, with a wind-down on Fridays, as indicated by their Jabber chats. This challenges

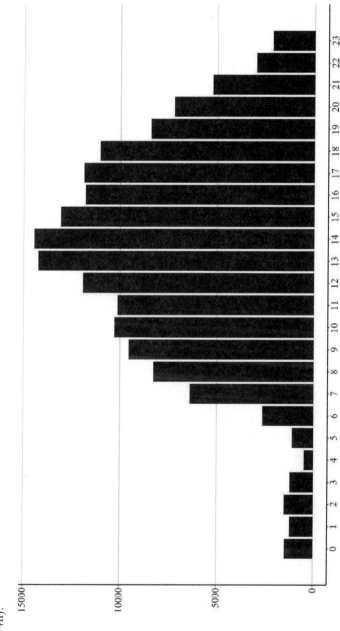

Figure 3: Jabber messages of Conti members distributed over the hours of the day. Most communication takes place at 2pm (14h).

a common belief that ransomware groups are busiest over the weekends. Yet it is evident that some members do choose to work while others rest. For example, the person tasked with posting victim data on Conti's leak sites prefers to operate outside of the conventional workweek.

The day-to-day experience for many Conti employees is not as thrilling or event-packed as one might expect. Much of their work involves routine, repetitive, and often mundane tasks, leading to a general dissatisfaction with their working conditions. Frustration is common among the staff, especially when they are expected to put in hours beyond the usual office time. One employee shared their perspective: "Like, it's up to you how you want to be. Here's my situation. I have eight hours of time. It was explained to me 50 for packages 50 for you, if you're not here in the morning, I don't sit around and wait, I just make packages and go through the process". This individual also expressed annoyance at having to handle tasks at times that disrupt their personal life, mentioning, "he throws in a task at 1 a.m. and says it needs to be done".

Recruitment and Onboarding

Conti cultivates a diverse team, drawing from a broad range of experience levels.[8] This is highlighted in a discussion where a senior developer mentions having four decades of experience. At the same time, Conti also opens its doors to those at the beginning of their technical journey in the field of ransomware, evident from Mango's commentary on the need for basic Linux skills and traffic monitoring capabilities: "you don't need super admins there, you need to know linux and be able to monitor traffic".[9]

Conti has put a systematic process is in place to recruit these people. HeadHunter, the leading Russian online recruitment service, operates by selling resume access and job advertisement space. Conti opts not to advertise its vacancies on HeadHunter to avoid drawing public attention. Instead, it bypasses the headhunter. ru platform, reaching out to prospects through direct emails. Conti is able to do this because they bought an unauthorized software package that provides access to the CV pool from HeadHunter.

While this approach enhances Conti's operational security, it also increases administrative burdens on the HR team, as there are a lot of unqualified applicants and they cannot access the site's internal filtering system.

Conti also taps into the underground market for recruitment, with Analyst1 researcher Jon DiMaggio pointing out postings by a figure known as "Khano"—likely identified within Conti's internal chats as Target—from September 2020 to February 2021.[10] These postings mention the search for long-term collaborators, promising "Cooperation on an ongoing basis", adding that "If we approach each other, you will get the most favorable conditions for work and creativity".[11] Furthermore, there is evidence to suggest that they may directly recruit students from universities. In a particular exchange, Stern inquires with Revers about preparing a test for candidates attending a university network event.

Conti recognizes that sometimes the most skilled hands are already busy at the keyboards of their competitors. This was evident in their strategy to attract talent from REvil, a former competitor. REvil once pulled off a dazzling stunt by depositing a million dollars in Bitcoin and then cleverly embedding a recruitment ad within a lively forum discussion about this deposit. This advert led to lots of public interest, with lots of aspiring ransomware operators leaving their contact information. Leveraging this, Conti's HR department sifted through the responses to compile a list of candidates to spam with job offers.[12]

When Conti's HR team identifies a candidate they believe fits a role within the organization, they anonymize the candidate's resume before passing it to a team leader. Checkpoint researchers highlight that this leads to a somewhat awkward back-and-forth, facilitated by HR, ensuring that the candidate's prospective team leader doesn't learn of their identity directly.[13] Nevertheless, this system is not without its flaws. On various occasions HR forgot to remove names from CVs. In another incident, a prospective recruit sidestepped conventional methods by breaking into the group's Jabber channel to contact Stern directly. Although this approach would typically be discouraged in regular corporate settings, within Conti, it was almost seen as a commendable achievement. Stern's response highlighted this attitude: he responded, "I respect you".[14]

There were individuals within Conti who were initially unaware that they had been employed by a criminal enterprise, instead believing they were part of a firm specializing in marketing, threat intelligence, or penetration testing. For instance, Silver, a senior member of Conti, misled a candidate about the nature of the business, claiming that the main focus was developing software for penetration testers: "everything here is anonymous, the main direction of the company is software for pentesters. the test task will be paid within a week (if not already paid), but most likely it will be within the next couple of days".

In a Jabber chat, Stern sheds light on their method of bringing these people on board. He says that a programmer often only gets deeply involved in a particular module, missing the bigger picture of the project. They need to invest significant time and effort to understand the overarching context of their role. Once they do, Conti offers them a pay raise. At this point, many employees interpret the smooth operation as indicating there are no negative consequences. Therefore, he points out, the choice to stay with or depart from Conti largely boils down to their moral compass.[15]

Compensation and Rewards

At Conti, there are two main categories of employees: those with a fixed base salary and those who earn commissions based on the organization's operational success. The latter category, typically reserved for senior staff, receives payments in Bitcoin, with annual earnings varying from $100,000 to $1.5 million.[16] Discussions about the percentage of commissions from ransomware activities are common, given their complexity. For example, Conti's use of botnets, such as Emotet and Trickbot, to enhance its ransomware campaigns affects the distribution of commissions. A leaked conversation shows Stern attempting to clarify the organization's roles and claiming a 20 percent profit share for himself.

The majority of Conti's workforce is on a base salary that aligns with the standard pay scales in Russia's industry. For instance, in August 2020, the HR department outlined the compensation plan for a new call center team, offering a fully remote job with work hours from 6am to 2pm Moscow time, Monday through Friday. The starting salary ranged from $450 to $500, with potential raises depending on the job role, and includes benefits such as paid vacation.

In his 2015 book *Swimming with Sharks: My Journey into the World of the Bankers*, investigative journalist Joris Luyendijk delved into the causes of the 2008 financial crisis and assessed the potential for its recurrence. He pinpointed the crisis's roots in harmful incentives within a competitive, male-dominated culture and declared the bonus-malus system in banking as fundamentally broken. Luyendijk argued for a substantial reform, suggesting that bankers should not only receive bonuses for success but also face maluses (penalties) for failures, advocating for increased personal accountability and risk-sharing.

Interestingly, the criminal ransomware enterprise Conti seems to have implemented this more effective incentive system, as a bonus *and* malus system appears to be in place—at least for the junior employees. In one Jabber exchange, senior manager Silver writes that three people were fined in a given month for "absenteeism and various mistakes that led to losses". These fines are then redirected into a pool designated for the "Employee of the Month" reward, a recognition granted by Silver on the last day of each month. The selected employee receives an additional 50 percent of their monthly salary. The management carefully selects the award recipient, often discussing the decision and providing explanations to the team. For example, in October 2021, the honor was jointly awarded to two employees: Collin and Ryan. Collin was recognized for successfully completing a "challenging project", while Ryan received the honor for taking the initiative on a new ransomware delivery method.[17]

Personal connections significantly influence the enforcement of disciplinary actions. For instance, a tester known as "Many" failed to report to work due to issues with his girlfriend and was subsequently dismissed by an irate middle manager. However, he appealed directly to Stern. In a long and emotional series of messages, he detailed the reasons for his absence and highlighted the lengthy period they had collaborated. Stern's response was concise: "work as you work".[18]

Friction, Confusion and Underperformance

Peter Drucker, the father of modern management theory, is often quoted as saying, "Only three things happen naturally in organizations: friction, confusion, and underperformance. Everything else requires leadership". This chapter's exploration into Conti's operations corroborates the observations made by various experts; there was a lack of leadership necessary to navigate Conti away from natural organizational pitfalls of friction, confusion, and underperformance.

However, it is important to contextualize these issues within the unique operational environment of Conti. Operating illegally, during a pandemic, amidst geopolitical instability, online, and anonymously exacerbate the inherent organizational challenges. Even the most skilled leaders would find it challenging to sustain an efficient ransomware enterprise.

8

BUSINESS EXPANSION

In early June 2021, Fortinet's Endpoint Detection and Response team, a cybersecurity company, stopped a ransomware attack aimed at one of their clients. Following a common pattern among ransomware families, this particular variant left a ransom note in text format in each affected folder. The ransom note contained a URL that directed victims to a ransomware website featuring a red banner with white text reading 'Diavol', apparently the name of the ransomware.[1] The choice of red in the website's design could be fittingly associated with the branding of the ransomware, considering that "Diavol" is the Romanian word for "Devil".[2]

The Fortinet team noticed that the deployment of the Diavol locker coincided with the appearance of locker64.dll on the victim's system, which was a later version of the Conti variant. While researchers observed similarities between the two locker files, there was not enough evidence to directly connect the two operations.[3] About a month later, IBM researchers discovered more concrete ties between Diavol and the Trickbot developers, the botnet group closely associated with Conti.[4] This discovery was further solidified when the FBI issued an advisory directly associating Diavol with the Trickbot group.[5]

The leaks shed more light on these connections. On August 19, 2021, Professor was outraged upon discovering a critical oversight:

the Diavol ransomware included a module from TrickBot designed to prevent attacks on CIS countries. This mistake provided security experts with the evidence needed to tie both TrickBot and Conti to the same cybercriminal entity. In his communication with the boss, Professor expressed his frustration: "Stern. Did you see how they fucked it up? Regarding the affiliation? I fucking almost exploded. They put a part of trickbot's code which is responsible for the reply on CIS into the build of Diavol. Despite me explicitly saying not to touch anything related to geo-targeting. And immediately the entire project is on the news as fully affiliated".[6]

The Diavol case offers valuable insights into Conti's evolution and branding strategies. It presents an intriguing scenario where Conti aimed to introduce a fresh product under a distinct brand that would not be immediately associated with their organization. This raises broader questions about Stern's and the senior leadership's business mentality, particularly in the context of expanding their criminal operations. Just like legitimate businesses, growth is a priority, and Conti is no different. But how they go about achieving that growth, and what tactics they employ to stay ahead of the competition, has significant implications for the cybercrime ecosystem. This chapter will delve into these growth questions, examining how Conti plans to expand its operation.

A framework often taught in business schools for analyzing businesses' growth and expansion strategies is the Ansoff Matrix, or the Product/Market Expansion Grid.[7] This model will be applied to dissect Stern's and Conti's expansion tactics in this context. The Ansoff Matrix simplifies the complexity of strategic options into a digestible format, plotting 'Products' against 'Markets' on two axes. It is designed to evaluate the appeal of different growth strategies by considering both the market and product dimensions, and the inherent risks of each approach. The matrix breaks down into four main growth strategies, creating a framework to understand how Conti might navigate their scaling ambitions.

Following this framework, this chapter begins with an examination of Conti's geographical expansion into Latin America. This move into new markets, along with the focus on targeting larger organizations capable of higher ransom payments in existing markets, demonstrates a calculated effort to boost profitability and expand influence. This chapter then looks into Conti's product development attempts to launch new ransomware strains under distinct brands and identities, highlighting the challenges in separating these new initiatives from the well-established Conti brand.

Table 4: The Ansoff Matrix

		Products	
		Existing	New
Markets	**Existing**	Market Penetration	Product Development
	New	Market Development	Diversification

At the heart of efforts to grow the business is Stern's significant push towards diversification. His ambitions were not limited to expanding Conti's reach or introducing new ransomware variations; he actively pursued broadening the group's portfolio of criminal activities. This included ventures into creating a new carding market and a social media platform, marking a shift towards a more varied criminal enterprise.

However, Stern's push to broaden Conti's range of activities might have inadvertently led to a neglect in their core product's innovation. The later versions of Conti's ransomware encryption mechanism, the 'locker', were noted for being less effective than their predecessors. This indicates a potential misstep in allocating too much focus on expansion and diversification at the expense of maintaining the quality of their primary 'product'.

Market Penetration and Development

Market penetration and market development are two of the most prevalent and lower-risk growth strategies in the business world. Market penetration aims to enhance the presence of current products within an existing market. This involves enticing customers from competitors and encouraging existing customers to make more frequent purchases. Strategies for achieving this may include reducing prices, boosting promotional and distribution efforts, acquiring competitors, or making modest product enhancements. If we take Uber as an example, this means increasing its user-base within an existing country or region for the same ride share services they already provide in the country or region. In contrast, market development involves identifying and developing new markets for existing products or services. This means increasing sales of existing products or services on previously unexplored markets. This strategy requires analyzing how a company's existing offerings can be introduced to new markets or how to expand within the current market. Approaches can vary; in the case of Uber, for example, it could be about exploring new geographic areas that previously didn't have the app, both domestically and internationally.

Conti has effectively employed both of these strategies. Its share of the ransomware market increased significantly between its first year of operations in 2020 and the subsequent year. While in the second quarter of 2020 it only held a small market share, some estimate that the Conti ransomware strain was responsible for around half of all ransomware infections in 2021.[8] The majority of their operations targeted organizations in the United States, followed by organizations based in Europe.

As the organization matured, its focus expanded not only in terms of persistence and breadth but also geographically. Data from the public leak site, Conti News, suggests that the group began to explore customers in markets beyond the Western world. In 2020, Conti only very sporadically released leaked data from victims in non-Western regions. For example, in early December 2020, they put data from a relatively small Indian company, called Ixsight Technologies, up for sale. They also offered data from a smaller information technology firm in the United Arab Emirates, known as CORE Information Technology Consultants. But these were exceptions to the rule.

However, a shift seems to have occurred in 2021. As shown on the two maps, Conti began expanding its operations into other

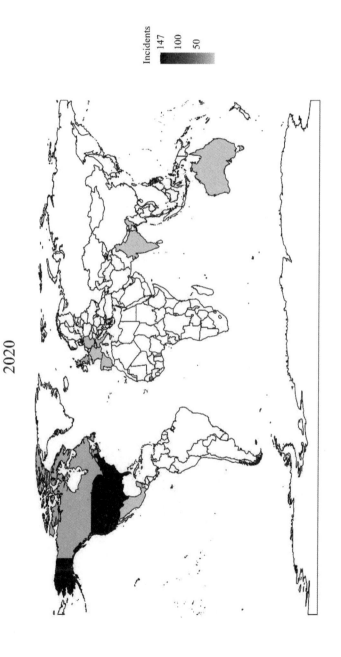

Figure 4: Number of Conti's self-reported ransomware attacks per country in 2020. There is a heavy focus on North America and Europe.

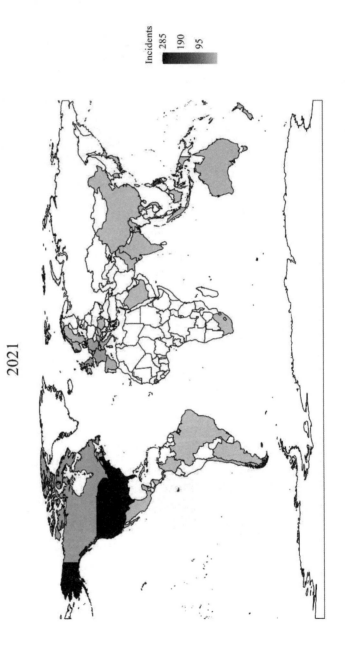

Figure 5: Number of Conti's self-reported ransomware attacks per country in 2021. Additional targets in Central and South America, as well as Asia, are included.

markets, with an emphasis on actively targeting organizations in Latin America. They listed the data of various Latin American companies on their leak site, including La Innovacion, a wholesale company in the Dominican Republic; Wenco, a plastics company in Chile; Fujioka Eletro Imagem, a retail business in Brazil; and Agricorp Company, a consumer goods company in Nicaragua. The most significant victim in terms of company size was Cable Color, a telecommunications company in El Salvador with over 10,000 employees. This trend of targeting Latin American organizations continued into 2022. Researchers from Recorded Future, a cyber threat intelligence company, posit that organizations in Latin America "may present as easy targets for ransomware attackers due to a general deficit of cyber resources, specifically education, hygiene and overall infrastructure".[9]

Over time, Conti increasingly targeted larger companies as well, shifting its focus towards customers capable of paying higher ransoms for their data. As Figure 6 shows, the group's interest in companies with only a few dozen employees significantly diminished, while it became more interested in those with over 500 employees.

Brand and Product Development

The aim of the product development strategy is to introduce novel products or services into existing markets. New products may be generated through various means: investing in additional product research and development, acquiring rights to manufacture external products, purchasing and rebranding products, or engaging in collaborative development with another company that seeks access to the firm's distribution channels or brands.

Conti implemented multiple product development strategies. The group sought to improve the pivotal component of its operations: the locker. As highlighted in the chapter discussing Conti's Playbook, the source code for Conti's locker version 3.0 was among the materials exposed by Danylo in February 2022. However, this new iteration was deemed a "giant step back from Version 2 in terms of code quality", as noted by various experts.[10] The modifications appeared to be the work of someone other than

Figure 6: Frequency of Conti attacks given company size in 2020, 2021, and 2022.

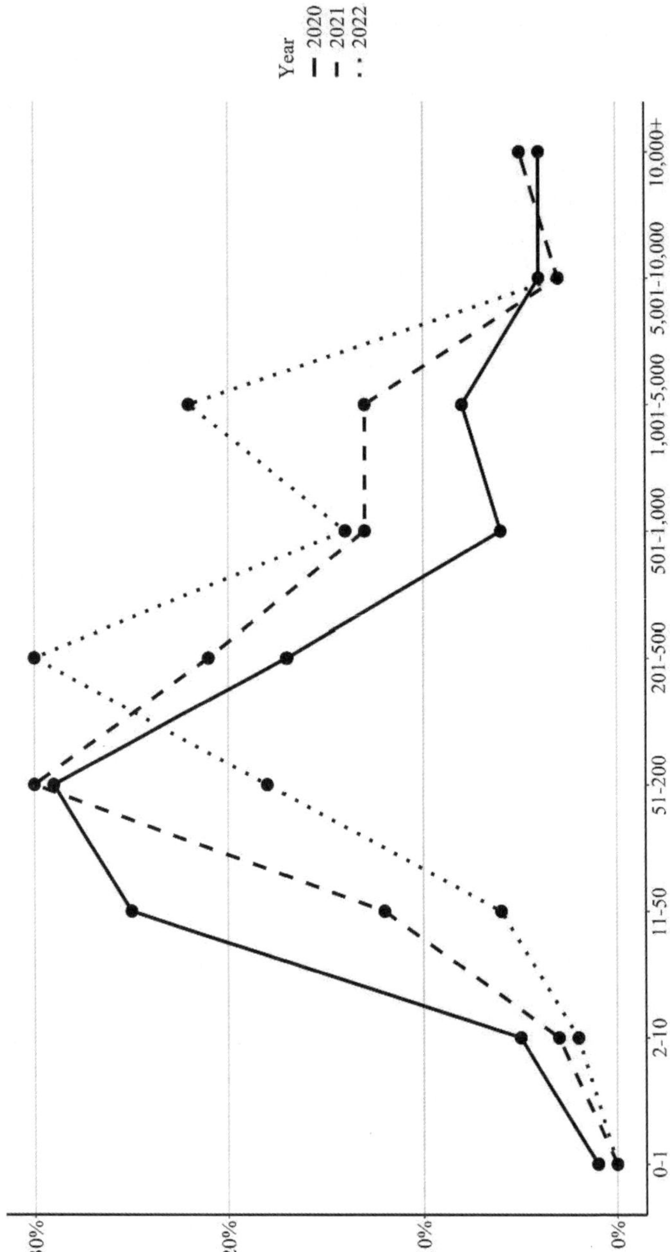

the original developer, lacking the same level of expertise. It is analogous to a beverage company attempting to refine their drink recipe but inadvertently worsening its taste, or a software company releasing a new version that is less user-friendly and plagued with more bugs.

The goal of Diavol was to create a brand that was distinct from the flagship brand. In chapter three on the MOB framework, I explained several reasons why a ransomware group may want to establish another brand. The exact motives behind Conti's actions remain uncertain, but based the angry messages from Professor, the most probable explanation is that leadership wanted to evade the watchful eyes of law enforcement agencies, threat intelligence companies and the media. They likely expected that Diavol would not initially attract the same level of attention, allowing the group to operate with a reduced risk of detection while conducting their ransomware activities. In other words, establishing Diavol was meant to provide a temporary shield against the scrutiny of public observers who track the criminal group's activities.[11]

Diavol was not the only new ransomware that Conti created. In the summer of 2021, a new group emerged under the name 'BlackByte'.[12] In the months that followed, Blackbyte went after several critical infrastructure sectors in the United States, including government facilities, financial services, and agricultural companies. The group faced a setback when Trustwave released a publicly available decryptor for BlackByte in October 2021.[13] Still, BlackByte showed resilience. They managed to fix the flaw, and their developers released newer versions that used multiple encryption keys, warning their victims against using the available decryptor on their leak site.[14] Their highest-profile attack came in February 2022, when they got access to files of the San Francisco 49ers, San Francisco Bay Area's professional American football team playing in the NFL. BlackByte stole personal information (including names and Social Security numbers) belonging to almost 21,000 individuals.[15] The computer systems were also likely encrypted, although the 49ers never confirmed this.[16]

Diavol and BlackByte can be seen as substitutes for Conti, offering comparable functionalities. Additionally, Conti developed a subsidiary element aimed at enhancing its primary activities. In June 2022, the FBI, CISA, and various other U.S. government bodies issued a joint advisory regarding a criminal group known as the Karakurt Hacking Team. This advisory highlighted that Karakurt, emerging a year prior, diverges from typical ransomware tactics by not encrypting stolen data.[17] Rather, they merely engage in data theft and pressure victims to comply with ransom demands through threats of publicly disclosing or auctioning the information. Essentially, Karakurt has shifted to a single extortion strategy, moving away from the more complex double or triple extortion tactics. This form of single extortion centers on the threat of releasing data rather than encrypting it. Despite this, the ransoms demanded by Karakurt remain very high, with amounts varying from $25,000 to as much as $13,000,000 in Bitcoin.[18]

Karakut derived its name from a venomous spider species, which they use for their branding and communication. On their leak site for victims, Karakurt writes: "Karakurts poison is very toxic and dangerous. Don't waste your time. What would you do? Of course you will have to take an antidote. In your situation it means that you still have a chance to survive. But it will cost as double. All you need is to accept our terms and conditions without any sort of bargain".[19] The payment deadlines are typically set to expire within a week of first contact of the victim.

Several security companies have released reports highlighting clear links between the infrastructure used by Karakurt and Conti. After acquiring access to a ProtonMail account used by Conti, researchers from InfinitumIT, a Turkish cyber security service provider, managed to infiltrate an internal Conti virtual private server (VPS) containing over 20 terabytes of stolen data. This server connection was subsequently traced to an IP address affiliated with Karakurt's leak site.[20] Arctic Wolf, a cybersecurity firm, investigated a re-extortion incident wherein Karakurt employed the same Cobalt Strike backdoor previously left behind by Conti in a prior attack on the same target. This finding strongly indicates that Karakurt had access to Conti's Cobalt Strike server, thus sharing operational infrastructure.[21] Additionally, Chainalysis presented further compelling evidence of a close relationship between the two groups looking at the payment process. They uncovered several Karakurt wallets that transferred cryptocurrency to wallets controlled by Conti, with victim payments ranging from \$45,000 to \$1 million.[22]

Vitali Kremez has provided a compelling analysis of Karakurt, positing it as a distinct brand under the umbrella of the Conti group, aimed at leveraging the shortcomings of Conti's ransomware operations.[23] According to Kremez, when Conti's ransomware deployment fails—specifically, if their encryption efforts are foiled and no data is encrypted—they turn to Karakurt. This subsidiary operation then uses the pre-exfiltrated data to extort ransoms from the victims, essentially capitalizing on the initial breach. This approach is further explained in a report by AdvIntel, the cybersecurity firm co-founded by Kremez, which details Conti's original operational model. As the report notes,

> [Conti's focus on encrypting data] was reflected in every aspect of their operations, including their reputation and technical infrastructure, which was designed to initiate negotiation with victims only after an encryption attempt was made. If the encryption tool, or "locker", failed to launch correctly, any stolen data would just languish on various storage platforms until it was eventually purged. Conti faced repeated situations where they successfully breached networks and stole data, but failed to monetize the operation due to technical issues with the locker execution.

To address this issue without major alterations to their operational model, Conti introduced Karakurt. This setup mirrors how a restaurant might repurpose unused ingredients from one day to create meals the next day, or how a fashion brand might use leftover materials to produce a less elaborate line under a new brand, effectively optimizing the use of resources and maintaining revenue streams even when primary operations do not proceed as efficiently as planned.

Diversification

Conti's leadership did not confine itself to the main business of ransomware but were actively seeking for new prospects to grow

125

the company.[24] Once we enter this domain of entrepreneurship, we encounter the fourth type of growth strategy: diversification. This strategy involves entering a new industry, while also creating a new product for that market. It is the riskiest approach among the four strategies, with the highest chances of failure. Nevertheless, it also promises many advantages. Not only can diversification help an organization to generate more profit and to gain a competitive advantage, but it is often also considered to be useful to avoid economic downturn as it spreads the organization's risk. In the context of ransomware, it can also be a helpful way to deal with legal downturn. When confronted with law enforcement pressure to reorganize and rebrand, having the opportunity to continue making money through other ways—that are not under investigation—is obviously quite beneficial in order to mitigate financial damage and make groups more flexible. Therefore, it is not surprising that Conti, a group wielding substantial purchasing power, initiated a quest to invest their capital in novel ventures and business development. The strategies devised by their leadership ranged from simple deceptions to fully-fledged side projects. Implementing these ideas, however, posed its own set of challenges, as the subsequent examples will demonstrate.

Stern had ambitious plans to create his own cryptocurrency. In late June 2021, he started discussing blockchain and cryptocurrency expertise with other members. He asked, "we want to build our own crypto system like: etherium, polkadot and binance smart chain etc. Does anyone know about it in detail? Study these above system, code, working principles. To build our own, where we can already stick NFT, DEFI, DEX and all the new trends that are and will be. So that others can create their own coins, exchanges and projects on our system". A few weeks later, Stern approached Logan, a member of the developer team, with plans for an alternative coin, often referred to as an altcoin—a catchall term for all cryptocurrencies other than Bitcoin—and the development of their own blockchain. He tasked Logan with exploring the necessary requirements, including the blockchain system, its code, and how it operates.

In the chapter discussing the operational playbook, I highlight that managing and laundering money is a critical issue for Conti.

Since ransomware payments are usually made in Bitcoin, and law enforcement has gotten better at tracking these transactions on the blockchain, Conti is considering creating its own altcoin.[25] This could help them avoid detection by law enforcement and security experts, giving them more control over their financial activities. This is particularly true if their altcoin includes privacy features similar to those found in the Monero cryptocurrency.[26]

Yet it is clear from the chat messages that this is not the main driver for Stern. He is not attempting to develop an altcoin to support his ransomware endeavors. Rather, he seeks to transition away from the ransomware business as he has grown weary of it. In a message from November 2021, during a period when Conti was at the height of its ransomware activities, he wrote: "I'm hovering now, interested in trading, defi, blockchain, new projects... There's an outlet everywhere. And I'm kind of bored with everything. I'm sure you'll get your own thing going there too. There's probably a big dream in this subject, I'm not sure it's necessary, but it will be useful for everybody. too many secrets big companies have, they hold on to, thinking it's their main value, these patents and data".[27] This shift in focus resembles a successful Silicon Valley startup founder who has amassed wealth and achieved business success but now yearns for new horizons.

Stern's frequent questions about the project and his willingness to invest a lot of resources highlight how important developing his own cryptocurrency was to him. Mango mentioned in an October 2021 message to channel members, "Blockchain needs people. So far it's all like a pervert's fantasy, but it's all real. and the boss seems to approve of all these expenses". Additionally, Stern's commitment included funding a $100,000 competition on the cybercrime platform Exploit, calling for various ideas for cryptocurrency platforms.[28]

By February 2022, Stern had put together a team to create a blockchain-based cryptocurrency platform, but they faced a massive challenge ahead.[29] Months earlier, Mango had predicted the difficulties, saying, "This is a great idea, but very complicated at the same time. Let's be realistic, we can't handle it on our own with so little experience and resources".[30] Stern often underestimated the effort needed to develop their own cryptocurrency platform, and Conti frequently found themselves stretched to their limits in terms of resources and expertise.

Stern's ambitious plans extended beyond cryptocurrency; he was also considering other business ventures. In May 2021, after the Colonial Pipeline ransomware attack, XSS, a prominent Russian-speaking hacking forum, announced a ban on discussions about "Ransomware affiliate programs", "Ransomware rental", and the "sale of lockers (ransomware software)".[31] This decision was made to avoid unwanted attention and to counteract the exaggerated focus on ransomware activities. Here is the translated text:

```
Degradation on the face. Newbies open up the media,
see some crazy virtual millions of dollars that they
will never get. They don't want anything, they don't
learn anything, they don't code anything, they just
don't even think, the whole essence of being comes
down to "encrypt—get $". They just run to github,
look for locker sorts there and run to encrypt every-
thing they see. Since our forum is aimed at begin-
ners, this factor is important to us.

Too much PR. Lockers (ransom) have accumulated a cri-
tical mass of nonsense, nonsense, hype, noise. When
you meet the "Ransomvarny negotiator" Profession,
you understand that you are in the looking glass or
just crazy. Moreover, 90% of this madness was cre-
ated artificially, feeding this hype. Those who make
good money on this noise (exchanges, insurance,
intermediaries, media, etc.)

Policy and hazard level. Peskov is forced to make
excuses in front of our overseas "friends"—this is
some kind of nonsense and exaggeration. The word
ranso was equated with a number of unpleasant phe-
nomena—geopolitics, extortion, government hacking.
This word has become dangerous and toxic.

Lockers will exist for a long time. This phenomenon
was too loudly promoted.
```

Criminal forums have a longstanding ambivalent relationship with ransomware. Before 2016, prior to the rise of ransomware as a service, reputable members of the Russian criminal community believed that deploying ransomware was not a cool thing to do. They viewed it as a squandering of botnet infrastructure and exploits, while some labeled it a "low-end tactic" leading to "intellectual stagnation".[32]

According to the leaked information, Stern directed Mango shortly after the XSS ban to explore the possibility of acquiring an existing exploit forum.[33] However, this effort ultimately failed, prompting Mango to propose to his superior the creation of their own social network. Mango elaborated on this idea:

```
Yesterday I got excited and fantasized few cool
features, such as personal file storage, with the
ability to share separate files, closed clubs based
on interests, entrance to which is a subject for
voting, public groups for general discussions. A
reputation system for users, so that we get rid of
grifters and intermediates and leave only the real
working guys … Here we can also add bunch of ser-
vices similar to what Babuk did last week, when they
allowed all hackers who want to post the data to
their own resources, even if they are on their own
and having nothing to do with their locker—anyone
can add any data there for storage.
```

Several months later, Mango presented two mockups to Stern created by a designer. The two also continued to brainstorm about what the platform should look like. Their ideas span from technical features like cryptocurrency integration to broader themes such as participant diversity and the overall design philosophy. Stern thinks that for the social network there should be "at least 1 million people", and that this should include "everyone": "reporters, regular users, buyers, hackers, and carders". The idea is to create a platform that has "all the shops in one place", with messaging, news, cryptocurrency integration, trading, and even gambling. However, he underscores that trading is the central pillar of the endeavor, and other features will follow. Later, Mango suggested a potential domain for the forum and a logo for it: "matryoshka[.] space (already with the domain:)) and as a logo matryoshka but angry, in our color scheme, dark-green, and may be draw a laptop next to it. In principle matches the theme. We are one big system amongst the multiple other sub-systems in one place. And it is clear that it is a Russian theme. It is going to be cool, and easy to remember, I think it will resonate with everyone".

The exuberant cryptocurrency market atmosphere at the time must have played a role in inspiring Stern's initiatives. In November 2021, Bitcoin prices had once again soared to all-time highs, surging past the $65,000 mark. Altcoins were also gaining a lot of traction, with even meme-driven tokens like Dogecoin capturing broad public interest. Equally, the discussion about the social forum seamlessly combing different aspects is reminiscent of the comprehensive ecosystem embodied by WeChat, a prominent Chinese app, and other trends we see in this space.[34] Notably, at time of writing Elon Musk is currently exploring ways in which his social media app X can incorporate additional features that encompass various facets of daily life.

Yet Stern's penchant for innovation did not solely focus on the future. Conti also looked back at the past, like their attempt to revive the carding market. As noted in the chapter on the evolution of ransomware, carding has been around since the late 1990s and was crucial for the criminal world. They were working on a new platform for this called "McDuckGroup".[35] The landing page of their website shows a picture of a Duck that is a mix of Scrooge McDuck and Darkwing Duck.[36]

Intentionality and Evolution

As Conti expanded and diversified its operations, the balance between strategic planning and adaptive evolution grew more complex. Stern's vision was ambitious, driven by a mix of business savvy, personal motivation, or the pursuit of better profits, and extended beyond ransomware to include various criminal activities such as cryptocurrency and social platforms.

It is crucial to recognize that Conti was not a unified entity but a collection of smaller groups, each with its own goals, skills, and level of independence. This structure probably shaped how the group handled branding and the creation of subsidiaries, indicating that changes in these areas might not have always been controlled by Stern or the main leaders. The launch of a new brand could indicate either growth or division within the group.

Furthermore, shifts in target selection might have been more decentralized. Rather than following specific orders from top management, individual operations might have adjusted their targets based on profit potential and existing network connections.

9

PIONEERING

In September 2021, Conti broke into the computer systems of
Graff, a high-end British jeweler catering to a wealthy clientele.
They extracted a vast amount of confidential information, such as
customer lists, receipts, invoices, and credit notes, and encrypted
various files of Graff's network.[1] Several weeks later, Conti
announced the successful compromise of Graff on their leak site,
providing links to 69,000 stolen documents that revealed details of
about 11,000 customers. They claimed this release represented
merely 1 percent of the total data extracted from Graff's systems.[2]

Nearly a month passed before *The Daily Mail* first covered the
incident and started revealing some of the names included in the
leaked data.[3] The list featured prominent figures, including politi-
cians like Donald Trump, Hollywood stars like Alec Baldwin or
Tom Hanks, and even royalty, like Bahrain's Crown Prince
Salman bin Hamad Al Khalifa and Saudi Arabia's Crown Prince
Mohammed bin Salman. The leaked information primarily consisted of addresses and receipts for jewelry purchases. By this
time, Graff announced that it had already been able to deal with
the encryption and restart its operations without any data loss, and
without paying Conti.[4]

However, after the first report by *The Daily Mail*, the story
quickly gained traction and was picked up by several newspapers,

bringing Graff and its clients lots of unwanted attention. Then, just a few days later, Conti removed the leaked data from their website and replaced the original message with something that can be described as a targeted apology along with a promise to do better. The new message from Conti first credits *The Daily Mail* for having done great journalistic work and in doing so having "uncovered things that we [Conti] have unfortunately missed" before publishing the sample data.[5] Conti goes on to state that they guarantee "that any information pertaining to members of Saudi Arabia, UAE, and Qatar families will be deleted without any exposure and review" and say that their team "apologizes to His Royal Highness Prince Mohammed bin Salman and any other members of the Royal Families whose names were mentioned in the publication for any inconvenience".[6] They further promise that the data was not sold anywhere else and that they will conduct their own review of it with the goal "to publish as much as Graff's information a possible regarding the financial declarations made by the US-UK-EU Neo-liberal plutocracy" and "raise awareness of the UK and EU governments who have regulations that legally prosecute the companies who can not protect their customer data".[7] Interest in the case gradually waned after this update and subsequent news pieces.[8]

Half a year later, it was revealed that Graff had complied with Conti's ransom demands, paying 118 Bitcoins, worth approximately $7.5 million at the time.[9] This payment came in mid-November, even though Graff had earlier stated that they had restored their systems. The Graff spokesperson justified the payment as follows: "we were determined to take all possible steps to protect their [the customers'] interests and so negotiated a payment which successfully neutralized that threat".[10]

The Graff incident sheds light on several facets of Conti's operations. It illustrates the potency of secondary extortion tactics, evidenced by Graff's decision to pay the ransom under heightened media scrutiny and after claiming system recovery. It also hints at Conti's procedure of reviewing data before leaks—albeit not meticulously. Furthermore, the case reflects Conti's preference for targeting entities in Western countries, avoiding engagement with or repercussions from non-Western entities. This avoidance could suggest a fear of retribution from powerful adversaries or a breach of their own ethical guidelines.

This raises a wider question regarding Conti's motives and ties to government entities. Conti's public message following the invasion in support of Russia, that led Danylo to leak the treasure trove of information, indicates at the very least that Conti's leaders back the Russian government and its actions. The critical question is: how far does this connection go? Are Conti activities directed or integrated by the Russian government? And how do these affiliations, if proven, intertwine with the group's operational methods? Are Conti's objectives aligned with state interests, or do they maintain a distinct agenda?

This chapter delves into the nuanced relationship between Conti and the Russian government, a relationship that resists being neatly categorized as "state-prohibited", "state-encouraged", or "state-ordered".[11] Conti operates with a certain degree of impunity within Russia, suggesting a tacit understanding with the authorities. Conti—like so many other ransomware groups—steers clear of targeting CIS countries. The situation around the Graff case highlights an extension of this non-aggression pact to include (senior members of) countries like the U.A.E.

Whilst Conti has membership outside of Russia, there is a pronounced pro-Russian and anti-Western sentiment, particularly against the United States. This ideological bent, combined with the lucrative opportunities in targeting American companies, positions the United States as a primary target for Conti's operations.

Conti operates autonomously and is not directly controlled by Russian state entities like the FSB. However, the group does engage in occasional collaborations with the government. These collaborations are initiated when the FSB calls upon Conti's hacking prowess for specific missions. Labelled 'pioneering', these initiatives are communicated through to Conti's upper echelon, with Stern also playing a pivotal role in these liaisons.

Conti's thus seems to engage in a calculated relationship with the Russian government. While the group harbors genuine pro-Russian sentiments and is willing to support the government in certain activities, it simultaneously prioritizes its independence and the pursuit of profit. This requires maintaining a connection with the government that is sufficiently close to avoid becoming a target of crackdowns, yet distant enough to prevent the government from gaining too intimate a knowledge of their operations and identities.

This chapter opens by discussing the unwritten pact to avoid targeting CIS countries and delves into Conti's tendency to focus on Western targets, with a particular emphasis on entities within the US. I then discuss the arrests of REvil members, an event that caused unease among some within Conti. Lastly, it examines the specific tasks that the Russian government, notably the FSB, assigns to Conti.

Profit Maximization and Geographical Interests

Conti's main goal is to maximize profits from ransomware activities, but this does not mean geopolitical factors do not play a role in its operations. As with many previous ransomware groups, Conti strictly avoids targeting countries within the Commonwealth of Independent States (CIS). Internal communications reveal that Conti members become extremely alarmed when they identify a Russian company or see the abbreviation 'OOO'—the equivalent of 'Limited Liability Company' in CIS countries—among their potential victims. This rule's importance is highlighted in various exchanges among Conti members. For example, there was an incident where Target and Troy realized they had inadvertently targeted what they believed to be a Russian company. Target vehemently cursed those responsible for the oversight, while Troy exclaimed, "they should be beaten up for such a thing. ... what if I did not notice that? Then we would have been fucked".[12]

Furthermore, there is a pronounced anti-Western, particularly anti-American sentiment within Conti, making the U.S. one of its primary targets. The U.S. also presents a large attack surface with numerous wealthy companies capable of paying significant ransoms. Additionally, as an English-speaking country, the U.S. simplifies many of Conti's operational processes, such as writing phishing emails, interpreting stolen data, and communicating with victims.

Following the further invasion of Ukraine in February 2022, Conti issued a statement on its leak site (which was later removed), declaring:

> As a response to Western warmongering and American threats to use cyber warfare against the citizens of Russian Federation, the Conti Team is officially announcing that we will use our full capacity to deliver retaliatory measures in case the Western warmongers attempt to target critical infrastructure in Russia or any Russian-speaking region of the world. We do not ally with any government and we condemn the ongoing war. However, since the West is known to wage its wars primarily by targeting civilians, we will use our resources in order to strike back if the well being and safety of peaceful citizens will be at stake due to American cyber aggression.

Internally, several members supported President Putin's rationale for the "Special military operation" in Ukraine. A member known as "Patrick" was particularly outspoken. On February 24, the day the conflict intensified, he posted, "war was inevitable, Ukraine made a bid for nuclear weapons". He added, "no one is happy about the war, brothers, but it is time to bring this neo-Nazi gang of Canaris's foster children to justice. […] Putin will address all questions today; I hope Kyiv will be ours by evening".

However, not everyone in the group saw the benefits of Russia's endeavors. Elijah expressed his confusion and disapproval in a chat, writing, "I never understood and still don't understand people who rejoice in war. Especially when your country is involved in this war". He followed up with a question to the group, asking, "Can you explain what's good about this?" The group is also international in scope, with various members in Ukraine. In reaction to Patrick's comments, a member named Kermit responded with a defiant, "Glory to Ukraine!"

The Arrests of a Peer-Competitor

By late 2021, Conti had firmly established itself in the ransomware ecosystem, having significantly expanded its operations. At that time, it faced few rivals, with REvil being a notable exception. However, REvil overreached with the July 2021 attack on Kaseya, an event detailed in chapter one. This attack resulted in ransomware infections across thousands of companies, attracting intense scrutiny from law enforcement and political figures. President Joe Biden issued a warning about taking "any necessary action" to defend U.S. infrastructure and pledged "the full resources of the government" to aid the investigation.[13]

In the wake of this incident, REvil became a target of heightened interest, likely leading to the sudden disappearance of their spokesperson, UNKN, and the group's decision to cease operations on July 13, leaving many victims stranded without a means to retrieve their data.[14] However, in a surprising development on July 22—nearly three weeks after the attack—Kaseya announced they had received a universal decryption key from a "trusted third party".[15]

It later emerged that the FBI was the "trusted third party" that had secured the decryption key during the REvil attack. They faced criticism for delaying the release of the key, which they defended as part of a broader strategy against REvil. Authorities were careful to avoid alerting REvil to their compromised network. The situation changed when REvil unexpectedly shut down, allowing the FBI to safely distribute the decryption key to Kaseya.[16]

However, REvil reappeared two months later, resuming their operations. Shortly afterwards, new data from victims began appearing on their leak site. With UNKN no longer present, a new spokesperson, known simply as REvil, took over communications.[17]

This comeback, however, was short-lived. By October 17, an REvil member known as 'o_neday' noticed unusual traffic redirection on their Tor site, which led to the discovery of a breach by an unknown entity. This discovery prompted a quick decision to shut down REvil's operations.[18] It was later revealed that the US Cyber Command was responsible for the hack.[19]

The international law enforcement community also stepped up their efforts against REvil. The US State Department added REvil to its bounty list, offering up to $10 million for information that could identify the group's leaders and $5 million for information leading to the arrest of its affiliates.[20] Through Operation GoldDust, a cooperative effort involving seventeen countries, along with Europol and Interpol, authorities arrested six affiliates and recovered approximately $6.5 million in ransom payments.[21]

The decisive blow to REvil came in January 2022. The Federal Security Service of the Russian Federation (FSB), in cooperation with the Ministry of Internal Affairs of Russia, announced the dismantling of REvil's illegal operations.[22] The FSB stated that "the basis for the search activities was the appeal of the competent US authorities, who reported on the leader of the criminal community and his involvement in encroachments on the information resources of foreign high-tech companies by introducing malicious software, encrypting information and extorting money for its decryption".[23]

The FSB reported that it had identified all members of the REvil criminal network and their involvement in illegal financial transactions. The Russian authorities also confiscated assets from twenty-five locations where fourteen members of the crime group lived. These assets included more than 426 million rubles, cryptocurrencies, $600,000, €500,000, computer equipment, crypto wallets used for criminal activities, and twenty luxury cars bought with the proceeds of their crimes.[24]

Many experts expressed skepticism regarding the arrests made by the Russian authorities. The authorities did not specify the number of individuals arrested or confirm if any leaders of the group were among them. Those detained were charged with violating Article 187, Part 2, of the Russian Criminal Code, concerning the "illegal circulation of means of payment".[25] This suggests that the arrests targeted not the key members responsible for the development of malware, but rather the lower-level money mules. Russia also expressed no interest in extraditing any Russian members of the group to the US.[26] Some analysts have speculated that the arrests might be aimed at recruiting these individuals to work for the FSB, possibly as a tactical move amidst related to war against Ukraine.

Despite this skepticism, the arrests impacted the criminal ecosystem. For one, REvil's collapse allowed Conti to expand, recruiting new talent and operators. These events also created anxiety in criminal forums, with concerns about Russia no longer being a secure base and eroding trust among ransomware groups. One analyst described the situation as "a major civil war going on [in] the Russian cyber-criminal underground".[27] For instance, the leader of Lockbit accused an individual known as "Kajit", the creator of a forum called RAMP that facilitated ransomware partnerships, of being a law enforcement agent or collaborating with the police.[28]

The REvil arrests also unsettled Conti members. Some struggled with the idea of FSB-U.S. collaboration.[29] Others blamed REvil's mistakes for their downfall.[30] Several Conti associates announced their departure from the ransomware business.[31] In response, internal discussions within Conti focused on devising strategies to prevent an outcome similar to REvil's. One employee, Angelo, believed Conti was safe due to Stern's close ties with the FSB and other Russian agencies (but not part of it).

* * *

Conti members are wary of extradition to the United States. Mango notes that "all the intelligence agencies in the world are looking for us", leading Conti to advise against international travel, especially to certain countries. As the Jabber leaks reveal, this advice was given to Skippy, an employee eager to travel for his ten-year wedding anniversary. While senior members cannot stop anyone from traveling, they stress not to take laptops and to wipe phones clean of sensitive information. Despite the risks, Skippy is keen to travel, citing others who have done so without issues.[32]

Conti members sometimes probe into each other's possible government ties. For example, Elroy asked Basil about any FSB affiliations, to which Basil cryptically responded: "I am not going to tell you where I am from (you understand that). But I have serious intelligence that on the border is not a training". This comment was made shortly before the further invasion of Ukraine in February 2022.

Basil then discussed REvil, noting that while their arrests made other group leaders nervous, law enforcement seemed to target lower-level members as they were charged with illegal money circulation rather than malware development. Despite being rivals, Elroy and Basil hope for the return of REvil's leadership and a new ransomware variant.

Orders from the FSB

Various experts have talked about more direct connections that tie Conti to the Russian government. In fact, on August 11, 2022, the U.S. State Department announced a reward up to $10 million for information on five Conti ransomware members. The announcement states: "If you have any information that ties hacking groups such as Conti, TrickbBot, Wizard Spider; the hackers known as "Tramp", "Dandis", "Professor", "Reshaev" or "Target"; or any malware or ransomware to a foreign government targeting U.S. critical infrastructure, you may be eligible for a reward". Accompanying the announcement was a potential photo of the Conti associate known as 'Target'. Notably, the document highlighted the reward for "foreign government-linked malicious cyber activity targeting U.S. Critical Infrastructure".

The leaks offer multiple examples that further confirm these close ties. Stern sometimes gets special requests from the FSB to carry out certain hacking operations. This activity is described as 'pioneering'. In this conversation from July 20, 2020, Stern and Professor discuss a government-directed task involving "academi". They confirm the assignment's origins—that "this is an office request, from one of the two" and specifics, focusing on accessing "chats". The discussion transitions from confirming the task to discussing compensation, with Stern downplaying the importance of immediate financial rewards in favor of long-term benefits. The conversation even includes a bit of humor, with Professor mentioning that he will "wear a red tie" whilst doing these activities.

Researchers from Trellix, a cybersecurity company formerly known as FireEye and McAfee, suggest that one of the two offices may be the famous 'Bolshoy Dom' (Big House), an office building at 4 Liteyny Avenue. This building functions as the central hub for the local branch of the FSB in Saint Petersburg, and it is mentioned in a later chat by another senior employee called Target.[33]

As the conversation continues, Stern and Professor appear to discuss the difficulty of a task and the importance of having the necessary permissions or "rights" to accomplish it. Stern also mentions that Target has plans to establish a new office to deal with government issues. They also discuss the global landscape, specifically mentioning the heightened government interest in COVID-19-related matters, and the activity of Cozy Bear—a designation used by CrowdStrike for a unit of the Russian Foreign Intelligence Service (SVR). This conversation took place shortly after a joint statement from the U.K., the U.S., and Canada, which condemned Cozy Bear's attempts to illicitly obtain COVID-19 vaccine and treatment research from academic and pharmaceutical entities worldwide.[34] The Kremlin, however, rejected these claims, stating that they were not backed by sufficient evidence.

A few months later, in a conversation from September 2020, Target mentioned that if Troy, another employee, succeeded in encrypting the data of Credit One Bank and eliminating all the backups, he would receive a reward from the Kremlin.

After the Conti leaks were published, Christo Grozev, the lead Russia investigator at Bellingcat, an investigative journalism group known for its fact-checking and open-source intelligence work, tweeted that his organization had been warned a year earlier. He was informed that "a global cyber crime group acting on an FSB order has hacked one of your contributors".[35] The hackers were searching for information on Alexey Navalny, the imprisoned Russian opposition leader. This came after Bellingcat had released several investigative pieces concerning the Novichok poisoning of Navalny.

As the Jabber chats show, in April 2021 Mango appears to be discussing this Bellingcat-related information with Professor about their interest in data linked to Bellingcat. Mango asks if their motivations are primarily patriotic rather than profit-driven in this operation, to which Professor confirms their patriotic commitment. Mango then raises concerns about potential decryption of the data they aim to acquire, indicating that he might trigger a beacon if decryption attempts are detected.[36]

During the conversation, Mango details his interaction with Johnyboy77, who provided information about a Bellingcat contributor's email and passwords relevant to Russia and Ukraine. "NAVALNI FSB" is mentioned, indicating a connection to the Russian security agency. Mango and Professor discuss the need for more specifics and the possibility of downloading files.[37]

A Delicate Balancing Act

Conti's operations suggest that there is a tension between their usefulness to the Russian government and their desire to remain independent. They seem to engage with the government when required but also seek enough distance to avoid being controlled. This careful balance allows them to operate without much interference, staying just close enough to gain some protection but far enough to maintain their freedom to sustain their lucrative operations.[38]

An interesting issue to consider is how Conti's move to physical office spaces might change this dynamic. Having actual offices makes them more exposed and potentially easier for the state to influence, especially since these locations can be directly accessed. This new setup could shift how Conti interacts with the state, possibly increasing the government's influence over their operations.

Moreover, the relationship between Stern—who effectively runs Conti—and the Russian security service, the FSB, is not entirely clear. This uncertainty adds another layer of complexity when assessing how deeply Conti is connected to state activities. Also, Conti is only known to maintain relationships with the FSB, not with the GRU, which is often linked to more destructive targeting. This distinction highlights Conti's focus, and perhaps limits the scope of their operations in terms of state collaboration.

10

DECLINE AND REVIVAL

Conti's cyberattack on the Costa Rican government in April 2022 signified a pivotal shift in the group's approach. Conti was unusually political and aggressive in their communication campaign through the messages on their leak site. This offensive against Costa Rica also marked Conti's final act of defiance. By May 19, 2022, all of Conti's online platforms had ceased operation, signaling the group's dissolution.

This chapter has two purposes. First, it delves into the complex series of events that precipitated Conti's collapse, with a focus on the repercussions of Danylo's disclosures and the decision to target the Costa Rican government. I analyze two differing perspectives to explain these events, starting with the 'Smokescreen Hypothesis'. This perspective suggests a calculated orchestration behind Conti's apparent fall, triggered by the leaks. In the time between these leaks and their ransomware attacks on Costa Rica, Conti was reportedly undergoing a covert restructuring. This involved enhancing their communication infrastructures and refining their tactical approaches. The aim was to maintain the illusion of the Conti brand's activity while gradually transitioning to new, lesser-known entities to avoid early detection. This maneuver, reflecting significant cunning and foresight, was chiefly orchestrated by the group's top management. Boguslavskiy most clearly articulated this viewpoint during his tenure at AdvIntel. The incident with Costa Rica is seen as a deliberate diversion, a masterful act of misdirection. I subsequently develop a second narrative, the 'Jumping off the Sinking Ship Hypothesis'. This suggests that the decline of Conti was already underway before the leaks emerged, with its leadership beginning to pivot towards other initiatives. In this context, the leaks served as a catalyst that accelerated the group's fragmentation. The unfolding events may not have been as intricately coordinated by the leadership as they appeared, particularly in light of numerous communication breakdowns. Although some factions within Conti might have attempted to create a smokescreen, the overall disorder and lack of a clear, cohesive strategy led to a more haphazard and uncoordinated departure by its members.

This chapter then discusses the criminal ecosystem that formed after the leaks and the fall of Conti. The scattering of employees, who sought new allegiances among various ransomware groups, fundamentally altered the dynamics and power structures within the ecosystem. As I will explain, the implosion of Conti birthed three new factions within the cybercriminal ecosystem. The first faction, 'Copycats', emerged from entities that often falsely claimed lineage to Conti. These groups either mimicked Conti's operations or developed further upon the Conti encryption practices, exploiting the leaked tools and tactics, yet without any direct organizational link to the original Conti group. The second faction, 'Retaliators', distinguished themselves from these copycats by their mission-oriented approach, utilizing Conti's tools *against* Russian interests in a bold counteroffensive.[1] The third faction, 'Offshoots', were groups comprising former Conti members who reorganized and continued their ransomware activities under new brands. These groups maintained the core organizational structures and playbook of Conti, albeit under a different identity.

The Smokescreen Hypothesis vs. the Jumping off the Sinking Ship Hypothesis

In a defiant message posted on the Russian-speaking underground forum RAMP on March 31, 2022, approximately a month after

DECLINE AND REVIVAL

Danylo publicly disclosed the leaked data from Conti, a member identified as 'Jordan Conti' emerged to set the record straight.[2] Known for providing updates on operations and recruitment, 'Jordan Conti' penned a vigorous defense of the group's resilience following the exposure:[3]

Ransomware philosophy post!

Gentlemen, before asking questions about how things are going with us, have you by any chance tried to look at our conti news website? It's kind of updated every day. EVERY DAY we leak more and more corporations, and these are only those that do not want to pay. And here the ratio is 2 to 1. That is, you can take the number of corps that we leaked, divided by two, and understand, approximately, how much we locked up with payment. And the average pay we have 700k. Practice entertaining mathematics, learn a lot of new things. And yes, if we can lock corporations every day and take their files on an industrial scale, then everything is fine with us, is it logical, right?

The locker is working and being finalized. With the leak, we lost exactly one asshole, who leaked the chats. And what? Do you really think that for a company of more than a hundred people, the loss of one foulbrood is the end?

This, by the way, can be said about every "drain". In the summer, the manuals were leaked, from which, O HORROR, the world learned that we were working through a koba! Yeah... this has never happened before, and here it is again! Then they leaked IPski, after which we put new ones! A death blow, of course... Then the chats were leaked, most of which were on a trick, which we already closed! Okay, now it's over for sure!

I'm for what, Conti is about like the "rotten west", which has been rotting among the Soviets for 100 years. "Now! Now it's definitely the end! The dollar will depreciate and there will be a hundred pieces per ruble, and also the external debt of the United States!" And somehow it turned out like this that there is no longer a scoop and no ruble, but for some reason everything is in order with the rotten west. Our group is like Kyiv,—it's already the second month since it was "taken in two days." And a lot of images can still be brought here, but the images are for pizdobolov (yes, the bum-infosec type from Twitter who reads this, I'm talking about you—we don't judge—bullshit helps you hide from yourself what you really are—a lose), but you and I need facts, right? And the facts are that we have been like a tank for four years in a row.

And there will only be more, believe me! New lockers, new penetration methods, new approaches to working with data.

In 2022, we open the "bending the world" achievement! And we are moving forward like a locomotive, and to all envious people—sail after that same warship—you are on your way there!

On that stood and will stand!

Jordan Conti invites skeptics to explore the Conti news site, which he asserts is thriving with daily updates, reflecting the group's continuous attacks on companies and other organizations that refuse to pay ransoms. Boasting a 50 percent payment success rate and an average ransom of $700,000, he emphasized that Conti was far from being subdued. The early April records on their leak site, listing almost a dozen new victims, supported his statements.[4] Additionally, data from Coveware indicates that the average ransom payout of $700,000 might indeed be accurate.[5] The operational status of their payment and 'support' portal also confirmed the group's active presence.

This portrayal of resilience prompted several threat intelligence analysts and cybersecurity experts to reconsider the narrative of Conti's decline. A report from Rapid7, a software company, is illustrative of these views: "Conti has repeatedly proven to be one of the most capable ransomware actors and these chats indicate that the group is well-organized and still very well-funded despite the schism. Any suggestion that these leaks spell the end for Conti is overstated, and we expect that Conti will continue to be a powerful player in the ransomware space".[6] Multiple other reports shared this perspective, reinforcing Conti's image as a formidable ransom war force, suggesting that the group remains cohesive and financially robust, despite any internal discord indicated by the leaks.

However, Boguslavskiy and Kremez from AdvIntel presented a contrasting view on these events, suggesting that Conti's apparent resilience was part of a more complex operation. Contrary to appearances, the maintenance of operations and the defiant messaging were components of a highly calculated smokescreen designed to obscure their actual situation following the leaks. Boguslavskiy and Kremez argued that shutting down and then attempting to resurrect a criminal brand as notorious as Conti's was an intricate process, one that could not be hastily executed without risking immediate detection and criticism. According to the researchers:

> A notorious and prolific threat group cannot simply turn off its servers, only to pop back up the following week with a new name and logo design. Even a whisper of novel threat group activity following the announcement of Conti's demise would likely spark immediate accusations of poorly executed identity theft. At best, immediate comparisons between the two would permanently leave the new group in Conti's ghostly shadow: the collective that fell and the one which emerged.[7]

In this context, the experiences of other ransomware groups like REvil and DarkSide, which also faced the challenge of disappearing and rebranding, become particularly illuminating. These cases highlight the difficulties and intricacies involved in either sustaining, dissolving, or transforming a ransom war group within the cybercrime landscape. Boguslavskiy and Kremez's interpretation suggests that Conti's response to the leaks demonstrates the group's deep understanding of these dynamics. Conti's actions and communications are not merely declarations of resilience, but maneuvers designed to obscure the true state of their circumstances and their future intentions: "As what was one of the dominant ransomware group active at the time, Conti realized that an element of *performativity* would need to be involved. Where other groups had been attempting a grand stunt with smoke and mirrors, Conti would try *a sleight of hand*".[8]

Over a period of more than two months, Conti was discreetly preparing for its transformation, setting up subdivisions that started operating even before the commencement of the group's shutdown process. These new subgroups either capitalized on the well-established reputation of Conti's existing aliases and locker malware or took the opportunity to establish new identities. This shift was consistent with Conti's historical approach of functioning through various subsidiaries under different brands.

The segmentation into smaller units, each still under the central command of Conti leadership, helped dodge law enforcement detection. It also provided a boost to smaller ransomware groups, now enriched with experienced Conti personnel who brought deep expertise in operations and negotiations. Among these new factions were specialized groups like Karakurt (discussed in chapter eight), which focused exclusively on data exfiltration, deviating from the traditional approach of data encryption.

In this intricate dance of deception and division, the remnants of Conti's infrastructure played an important role. According to Boguslavskiy and Kremez, "Conti's remaining infrastructure operated like an army preparing for an ambush. Lingering actors were left to keep their fires lit, visible from behind enemy lines. Meanwhile, hidden from view, Conti's most skilled agents were instead laid low in a nearby encampment, biding their time while watching their great and empty camp send out smoke signals, meticulously emulating the movements of an active group".[9] Many entries on Conti's leak site were, in fact, from older breaches.[10] This approach enabled Conti to uphold an illusion of resilience and growth, demonstrated by continued leaks of documents from prior breaches and aggressive self-promotion in criminal forums.

From this perspective, Jordan Conti's proclamation of the group's unwavering success and the recruitment of new affiliates were all part of a wider ruse, crafted to divert attention and speculation away from the group's covert activities. The high-profile attack on the Costa Rican government epitomized this tactic, serving both as a distraction and as proof of Conti's lingering power. This elaborate display was not just about reinforcing Conti's legacy; it was designed to ensure a smooth transition for its offshoots into their next criminal phases, all under the guise of ongoing operational vitality and strength.

* * *

I would also like to offer a different interpretation of events, presenting Conti as an organization already grappling with significant internal challenges long before the leaks came to light. I discuss this perspective as the 'Jumping off the Sinking Ship Hypothesis'. This view posits that the operational difficulties and leadership rifts Conti faced were not simply the result of external pressures but were deeply embedded within its structure. A key factor in these issues was the reliance on a pivotal leader, Stern, whose withdrawal marked the beginning of Conti's decline.

The internal communications shed light on the severity of the situation, particularly through a message from a member known as 'Frances'. Dated February 21, 2022—before the further invasion of Ukraine by Russia and the exposure of Conti's messages— Frances' message to the collective underscores the critical nature of their situation:

Friends!

sincerely apologize for the fact that the last few days I was forced to ignore your questions. Regarding the Chief, Silver, sn and everything else.

Forced due to the fact that I simply had nothing to tell you. I pulled the rubber, got out of the RFP as best I could, hoping that the boss would appear and clarify our further actions.

But there is no boss, and the situation around us does not become softer and I no longer see the point in pulling the cat by the balls.

We have a difficult situation, too close attention to the company from the outside has led to the fact that the chief apparently decided to lay low.

There have been many leaks, there have been post-New Year's parties and many other circumstances that are tempting us to take a little vacation for all of us and wait until the situation settles down.

The reserve money, which was set aside for emergencies and urgent needs of the team, was not even enough to close the last salary payment. There is no boss, there is no clarity and certainty with further affairs, there is no money either.

We hope that the boss will appear and the company will continue to work, but for now, on behalf of the company, I apologize to all of you and ask you to be patient. All balances on the RFP will be paid, the only question is when.

Now I'll ask you all to write to me in a personal: (ideally in a toad:))
- Actual backup contact for communication (it is desirable to register a fresh public toad
- Briefly your job responsibilities, projects, PL (for coders). Who did what, literally in a nutshell.

In the near future, we, with those team leaders who remained in the ranks, will think about how to restart all work processes, where to find money for salary payments and launch all our work projects with renewed vigor.

As soon as there is any news on payments, reorganization and return to work, I will contact everyone. In the meantime, I have to ask all of you to take a 2-3 month vacation.

We'll try to get back to work as soon as possible. From you—we ask everyone to take care of your personal safety! Clean up working systems, change accounts on forums, VPNs, if you need phones and PCs. Your safety is your first responsibility! In front of yourself, in front of loved ones and in front of the team too!

I ask you not to break the PM with questions about the boss—I won't tell anyone anything new, because I simply don't know.

Once again I apologize Friends, I myself am not enthusiastic about all these events, we will try to somehow correct the situation.

> Those who do not want to move on with us—we natu-
> rally understand. For those who will wait—we rest for
> 2-3 months, take care of our personal lives and
> enjoy freedom:)
>
> All working rockets and internal logs will soon be
> disabled, further communication—only on backup logs.
> Peace for everyone!

This message reveals the precarious state of Conti, acknowledging a significant internal crisis. Frances' message called for team members to brace for a hiatus, secure their operational footprints, and await further directives—a clear contrast from the resilient facade Conti maintained in public forums.

On the same day that Frances communicated his concerns on the forum, Mango also sent a message to the team, attempting to manage expectations and provide some reassurance amid the escalating uncertainty. Mango admitted that the situation was "a mess" but remained hopeful, suggesting that "within the next two weeks everything will be sorted out". This statement captures the chaos enveloping Conti even *before* the intensified conflict in Ukraine and Danylo's leaks, depicting the organization as deeply troubled.

Strix, inquiring about new projects and pending financial compensations, received an evasive response from Mango, who indicated that no new assignments would be forthcoming for the time being. Mango outlined two possible solutions: either Stern's return would resolve their issues or, failing that, Mango and Defender would need to devise an alternative plan. This dialogue underscores the group's reliance on Stern for critical decisions and the challenges of continuity in his absence. Mango tried to offer reassurance to Strix, saying, "we'll pay everyone back".

In truth, Defender had already expressed concerns about Stern's absence since late 2021, as indicated in various internal communications: "hello, when are you going to drop off the work and pay? I'm paying for servers out of my own already". Defender recommends to Stern to just transfer $10,000 so he does not have to distract him too often. He later asked again "hello, you tell me, why are you ignoring me in terms of salary? I'm already overdue on my credit, no money on my job, I need to renew servers today".

The leaked messages underscored a growing financial instability within Conti, with critical operational expenses and salaries remaining unpaid. Stern's inattention to these obligations suggests a leadership disengaged from the group's day-to-day realities, a situation that inevitably led to discontent and uncertainty among the ranks. Figure 7 shows the number of messages sent by Stern via Jabber. It shows that he became inactive from late 2021. The UK National Crime Agency (NCA) has reported that the cryptocurrency wallet linked to Stern held an estimated $95 million at around the time he vanished.[11]

Thus, the decision to target the Costa Rican government can be interpreted not as a calculated maneuver but as a desperate act by a faction within Conti struggling to assert its relevance and capability in the absence of coherent leadership and financial stability. With the core of its leadership either disinterested or absent, and financial resources dwindling, Conti was a shadow of its former self, its actions driven more by desperation than by any strategic foresight. This narrative challenges the notion of Conti's calculated resilience, suggesting instead a fragmented group scrambling to navigate its own disintegration.

The 'Jumping off the Sinking Ship Hypothesis' does not rule out the possibility that some members of Conti consciously regrouped or that certain attacks were tactically executed to maneuver through challenging circumstances. Rather, it casts doubt on two key points: firstly, the extent to which the leaks directly precipitated Conti's downfall, and secondly, the nature of the calculated maneuvers purportedly made in the aftermath of these leaks. It suggests a more complex narrative where internal fragilities and external pressures interplay, questioning the narrative of a cohesive, strategic response to the leaks.

Copycats, Retaliators, and Offshoots

The exposure of Conti's operational blueprints by Danylo is a double-edged sword. The leaks served as a crucial resource for security research, enhancing defensive measures against ransomware threats. As elaborated in earlier chapters, they improved the

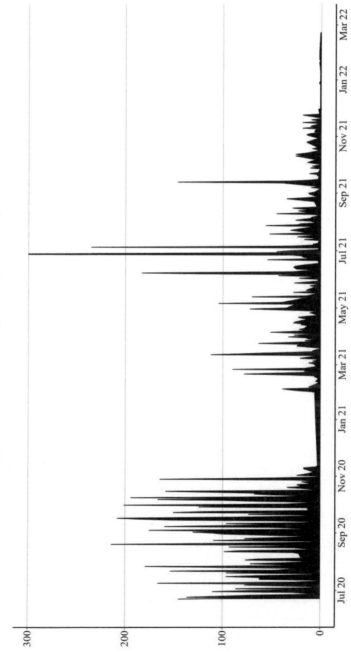

Figure 7: Number of messages sent by Stern on Jabber

precision in tracking Conti-related cryptocurrency wallets and their transactions through various exchanges. Additionally, the leaks shed light on the group's internal hierarchy and upcoming ventures, and confirmed suspicions of state affiliations, offering solid proof where there was once only speculation. They also exposed Conti's practice of deploying multiple brands to mask their operations' interconnectedness, thereby enriching our understanding of the group's extensive operations, revealing intricate details about their organizational structure and strategic maneuvers.

At the same time, the leaks equipped a wider circle of cybercriminals with reliable tools and methodologies previously exclusive to Conti. Although Danylo did not share the password of Conti locker's code base publicly, it was not long before another researcher deciphered it, granting widespread access to an exceptionally refined and clean codebase. This development encouraged criminals to either replicate or enhance these established methodologies.[12]

Various copycat groups emerged, either imitating Conti's operations or building upon its encryption techniques using leaked tools and strategies, yet without any direct organizational ties to the original group. For instance, ransomware groups such as Scarecrow, Meow, and Putin leveraged the leaked Conti v2 ransomware builder.[13] All three groups communicate with their victims and publish victim details using Telegram channels. In the second half of 2022, BlueSky ransomware also appeared, exhibiting several similarities with both Conti and Babuk's source code.[14] Upon activation, BlueSky ransomware encrypts files, appending the .BLUESKY extension, and drops a ransom note titled "# DECRYPT FILES BLUESKY #.txt" with instructions for decryption. This group manages victim interactions through an onion site, only accessible via Tor.

More established groups have also adopted Conti's source code. The LockBit ransomware gang, the most widespread post-Conti, integrated encryptors based on Conti's leaked source code. LockBit evolved through several versions to LockBit 3.0, also known as LockBit Black,[15] and introduced a version based on Conti's source code named LockBit Green. The adoption of Conti's source code by LockBit suggests that former Conti operatives may have transitioned to LockBit. According to researchers from PRODRAFT, "We especially observed that ex-Conti members preferred LockBit Green after the announcement. They likely find comfort in using Conti-based ransomware".

The most prominent Conti copycat appeared a few months after Conti's attacks on the Costa Rican government. Organizations infected by this new strain received a ransom note virtually identical to Conti's, with two small but significant changes: "CONTI" was replaced with "MONTI", and a new Tor-based (.onion) URL was provided for contact.[16] The mimicry was so exact that the note included, "If you don't know who we are—just 'Google it'". This was ironic, given Monti's absence from public record, with no media or threat reports on them as of June 2022, except for a mention in a tweet from the MalwareHunterTeam.[17]

Monti's attempt to replicate Conti's tactics and success was marred by amateurism.[18] Detailed analysis by researchers from Blackberry indicated that as of July 5, 2022, the .onion domain meant for contacting Monti was inaccessible, with no evidence that it had ever been operational. Public and darknet research, and communications with incident response firms, found no proof that the domain ever functioned. The lack of evidence from other Monti cases suggests we may never confirm if the domain was accessible. Further analysis of the leaked executable associated with Monti showed that, although the code was identical to Conti's, executing the Monti payload did not result in any file encryption, indicating a failure in operational execution.[19]

* * *

The second faction, 'Retaliators', set themselves apart from these copycats by adopting a mission-oriented approach that used Conti's tools specifically *against* Russian interests in a bold counteroffensive. When Russia escalated its military activities in Ukraine in February 2022, various new groups announced on X their intent to digitally support Ukraine. Among them was Network Battalion 65 (NB65), which declared that they only targeted Russian entities.[20] Their first reported victim was on February 27, 2022, attacking Fornovogas, a Russian energy company, where they compromised control systems, causing disruptions like gas leaks, fan shutdowns, and deletions of profiles.[21] That same day, they claimed to have stolen 40,000 files from the Russian Nuclear Safety Institute in Moscow.[22]

In subsequent weeks, NB65 escalated their activities, claiming attacks on several high-profile Russian organizations including the manipulation of SCADA control systems at Severnaya Kompaniya, server shutdowns and credential tampering at the Russian Space Research Institute (Roscosmos), and the exfiltration of 870 GB of data from the All-Russian State Television and Radio Broadcasting Company (VGTRK).[23] They even accessed pre-recorded videos from The State Technological University.[24] The cybersecurity community began taking serious note of NB65 when they claimed to have hacked Kaspersky Labs and threatened to leak its source code, but then, nothing of importance was actually leaked.[25]

However, by late March, NB65's tactics shifted. They targeted Mosexpertiza, the Moscow Independent Center for Expertise and Certification, encrypting data with Conti's locker and stating, "there will be no negotiations", opting instead to share the data with DDoSecrets, a platform they used in previous operations. This marked a departure from their earlier disruptive tactics.[26] There was no attempt at extortion, but this changed in subsequent operations. In early April, they encrypted data and deleted backups at SSK Gazregion LLC, offering decryption if the company complied with the instructions in a readme.txt file they left.[27] They continued this pattern a few days later, on April 6, targeting Continent Express, a privately-owned company.[28] This action was justified by events in Bucha, prompting NB65 to extend their targets to include private businesses, with the promise of decryption if instructions were followed.[29] On April 18, NB65 targeted the Petersburg Social Commercial Bank (JSC Bank PSCB), mocking them by tracking and posting their internal communications on X, highlighting the bank's lack of response.[30] NB65 half-jokingly claimed to have modified Conti's source code "to make it more effective against Russian targets". A month later, they breached the bank again, ridiculing them for not patching their vulnerabilities and stealing more data, declaring again that there would be no negotiations. NB65's ransomware operations continued for several weeks, with claims of "vastly improving" the Conti encryption mechanism over time.[31]

A spokesperson for the NB65 hacking group told *Bleeping Computer* that they adapted their encryptor from the initial Conti source code, tailoring it for each victim to ensure that existing decryptors would not work. "It's been modified in a way that all versions of Conti's decryptor won't work. Each deployment generates a randomized key based off of a couple variables that we change for each target", NB65 explained to *Bleeping Computer*. "There's really no way to decrypt without making contact with us". An analysis of one sample uploaded to VirusTotal showed that NB65's ransomware used 66 percent of the same code as typical Conti ransomware samples.[32] When executing its ransomware, NB65 appends the NB65 extension to the names of encrypted files.[33]

The ransom note from NB65's ransomware credits Conti for the technical tools and blames the cyberattack on President Vladimir Putin for his actions in Ukraine:

NB65

By now it's probably painfully apparent that your environment has been infected with ransomware. You can thank Conti for that.

We've modified the code in a way that will prevent you from decrypting it with their decryptor.

We've exfiltrated a significant amount of data including private emails, financial information, contacts, etc.

Now, if you wish to contact us in order to save your files from permanent encryption you can do so by emailing network_battallion_0065@riseup.net.

You have 3 days to establish contact. Failing to do so will result in that data remaining permenantly encrypted.

While we have very little sympathy for the situation you find youserlves in right now, we will honor our agreement to restore your files across the affected environment once contact is established and payment is made. Unitl that time we will take no action. Be aware that we have compromised your entire network.

We're watching very closely. Your President should
not have commited war crimes. If you're searching
for someone to blame for your current situation look
no further than Vladimir Putin.

On August 1, 2022, NB65 tweeted a message that seemed to indicate internal challenges or threats: "People want to play stupid games they're going to win stupid prizes. You know who you are. Working against us means you're working for the enemy. You fucked up so bad. You're damaging our ability to operate, and that means you're helping the enemy. Eat fucking shit". Then, their last tweet appeared on the same day, containing an encrypted message that translated to the following: "We love our Ukrainian brothers and sisters. Network Battalion 65 are removing ourselves from the public eyes of social media. We will be gone but we will not stop fighting the war. Russia will still feel the sting of our attacks, and we will continue to support Ukrainian cyber operations for as long as it takes. Thank you for the vast amounts of love and support from our followers. Slava Ukraini! Fuck Russia Forever... Always Watching, Always Fighting, NB65 >3". Since then, nothing has been heard of or reported on them.

NB65 likely started as a small group or even an individual hacker motivated by Russia's invasion of Ukraine, and it later attracted more members. After initially just extracting and leaking data, two months post-Conti leaks, they began using a modified version of the Conti encryption tool and initiated a sort of double extortion scheme. However, the extortion attempts appear to have been largely unsuccessful, with no known victims contacting NB65. That said, extortion was never their primary objective.

The third faction, 'Offshoots', consists of former Conti members who transitioned to established brands that existed during Conti's active period, such as Blackbyte and Karakurt—discussed in chapter seven—as well as those who have reorganized and continued their ransomware activities under new brands.[34] Due to the absence of leaks similar to those from Conti and the challenges in verifying potential human intelligence (HUMINT) from commercial threat intelligence sources, it is difficult to confirm details about the internal dynamics of these groups. However, several groups are noteworthy.[35]

A prominent group believed to have evolved from Conti is Black Basta. First observed in April 2022, evidence indicates that this RaaS group had been in development since at least February 2022.[36] Black Basta is particularly known for its focus on big game hunting, targeting large organizations in the construction, manufacturing and healthcare sector.[37]

In its initial two weeks alone, Black Basta listed at least twenty victims on its leak site, known as Basta News. The rapid accumulation of high-profile victims indicates a highly organized ransomware group that likely did not emerge spontaneously. Many researchers have observed similarities between Black Basta and Conti in terms of TTPs, data leak site infrastructure, and victim recovery portals, although they are cautious about asserting the extent to which former key members of Conti are involved with Black Basta.[38] Microsoft reported that one affiliate, identified as DEV-0506, "was deploying BlackBasta part-time before the Conti shutdown and is now deploying it regularly".[39]

Another group that emerged in the wake of Conti is known as Silent Ransom, also referred to as Luna Moth. This group specializes in callback phishing extortion campaigns targeting various sectors, including legal and retail. Callback phishing, or telephone-oriented attack delivery (TOAD), is a type of social engineering that requires direct interaction between the threat actor and the target to achieve their goals. This method, while more resource-intensive, is simpler than script-based attacks and tends to have a higher success rate. Historically, groups associated with the Conti group have successfully used this attack method in the "BazarCall campaign".[40] AdvIntel believes Silent Ransom has ties to Conti,[41] though some sources are cautious about confirming this connection.[42]

Kristopher Russo from Unit42 outlines a typical callback phishing attack chain from Luna Moth initiated by a phishing email to a corporate address, featuring a seemingly minor fraudulent charge on an attached invoice (typically under $1,000) which, if unnoticed, allows attackers to evade email protection systems.[43] Recipients who call the phone number provided are connected to a threat actor-controlled call center, where they are tricked into downloading a remote support tool under the guise of canceling a subscription. This enables the attackers to install a remote administration tool or directly access and exfiltrate sensitive data if administrative rights are absent. The stolen data is then used to extort a ransom from the victim, with threats of public disclosure to customers and clients if payment is not made, escalating demands if the victim does not engage.

The final notable offshoot group is Royal Ransomware, by some considered "the direct heir of Conti".[44] Royal appeared in early 2022, initially using third-party ransomware like BlackCat and custom Zeon (also connected to Conti) before transitioning to its 'proprietary ransomware' since September 2022.[45] It operates independently without adopting the RaaS model.

Royal initiates attacks through phishing campaigns utilizing common loaders such as BATLOADER and Qakbot. The ransomware is written in C/C++ and executed via command line, indicating sophisticated, human-led breaches that involve penetration testing teams who facilitate network access, privilege escalation, and lateral movement.

Researchers from Cybereason note that Royal's unique approach to encryption. Unlike typical ransomware that partially encrypts files based purely on size, Royal allows operators to adjust the percentage of the file encrypted, enhancing evasion capabilities even with larger files. This method is similar to Conti's strategy, where files over 5.24 MB are encrypted at 50 percent in a segmented manner.[36]

A group that likely later emerged from Royal is ThreeAM, also spelled 3AM.[47] An interesting tactic used by this group is their method of increasing pressure on victims by automating replies on X, where they broadcast news of their successful attacks. They post under various news posts, mentioning one of their victims and directing readers to the data leak site.[48]

Table 5: Overview of Conti related groups[49]

Date	Event	Group
Aug 2017	Hermes ransomware available for purchase on Exploit.in	Hermes

Aug 2018	First appearance of Ryuk	Ryuk
Late 2019	Ryuk increasingly relies on Emotet and Trickbot	Ryuk
Late 2019	Ryuk becomes most used ransomware strain	Ryuk
Dec 2019	First appearance of Conti	Conti
Apr 2020	Decline of Ryuk, rise of Conti	Ryuk
Jun 2021	Karakurt starts registering domains and accounts.	Karakurt
Jul 2021	First appearance of BlackByte	BlackByte
Jul 2021	Diavol and Conti released in same ransomware operation	Diavol
Aug 2021	Conti training manuals leaked by M1Geelka	Conti
Sep 2021	First appearance of Karakurt	Karakurt
Sep 2021	Quantum emerged as rebranded Mountlocker	Quantum
Oct 2021	Free decryptor Blackbyte released due to flaw	BlackByte
Jan 2022	First appearance of Zeon	Zeon
Jan 2022	First appearance of Royal, using third-party ransomware	Royal
Feb 2022	(Likely) start development Black Basta	Black Basta
Mar 2022	First appearance of Silent Ransom / Luna Moth	Silent Ransom
Apr 2022	Conti's attack on Costa Rica	Conti
Apr 2022	First appearance Black Basta	Black Basta

May 2022	Conti's leak site terminated	Conti
Jun 2022	First appearance of Monti	Monti
Jun 2022	Quantum uses Bazarcall in a Jörmungandr campaign	Quantum
Jun 2022	Roy/Zeon use of Bazarcall	Roy / Zeon
Aug 2022	New version of Blackbyte released	BlackByte
Sep 2022	New versions of Diavol uploaded to VirusTotal	Diavol
Sep 2022	Royal starts using its own ransomware	Royal
Oct 2022	First appearance of Bluesky	Bluesky
Oct 2022	Version of Scarecrow released with Conti elements	Scarecrow
Mar 2023	First appearance of Akira	Akira
Sep 2023	First appearance of 3AM	3AM group

An Evolving Criminal Ecosystem

In 2021, the ransomware ecosystem was marked by significant monopolization. However, post-2022 witnessed a dramatic shift towards fragmentation and decentralization.[50] The landscape now comprises of many more groups, each specializing in different aspects of ransomware operations. Some manage the complete cycle of an attack—from intrusion to extortion to payout—while others provide specific tools and services, delegating the execution to different actors. This diversification has added layers of complexity to understanding the ransomware ecosystem, as groups continually realign and evolve.

The emergence of these offshoots from early 2022, along with the strengthening of earlier brands, supports both the 'Smokescreen' and 'Jumping off the Sinking Ship' hypotheses. The 'Smokescreen

Hypothesis' suggests that the downfall of Conti and its actions in Costa Rica were part of a well-orchestrated rebranding effort. Conversely, the '*Jumping off the Sinking Ship* Hypothesis' aligns with the view that Conti was already in decline, with the leaks merely accelerating its demise.

Yet, the two narratives offer different perspectives regarding the motivations and structures that have emerged post-Conti. The '*Smokescreen* Hypothesis' posits a deliberate selection of brands and a centralized system that is paradoxically decentralized by design, indicating a top-down strategic approach. Boguslavskiy further explains to *Bleeping Computer* that rather than rebranding as another large ransomware operation, Conti leadership has partnered with smaller ransomware gangs.[51] This partnership provides smaller gangs access to a wealth of expertise from seasoned Conti pentesters, negotiators, and operators, enhancing their capabilities. Simultaneously, the Conti syndicate achieves increased agility and better evades law enforcement by distributing its operations across smaller, discrete "cells" that are coordinated by a central leadership.[52]

In contrast, the '*Jumping off the Sinking Ship* Hypothesis' suggests a more grassroots, bottom-up approach. This perspective shows a variety of new groups, some maintaining close ties to Conti's original leadership like Stern, while others significantly diverge.

Despite these two narratives, it is important to recognize that the landscape continues to evolve. We already observe the emergence of various other Conti-related groups like 3AM and Akira. Over time, as these entities mix with other ransomware groups and strains, it becomes increasingly challenging to trace clear connections back to the original Conti group.

CONCLUSION

Governments have certainly made efforts to prevent, deter, and disrupt Conti's operations.[1] On February 4, 2021, U.S. authorities struck an early blow by arresting Alla Witte, a developer part of Conti and Trickbot, in Suriname.[2] A few months later, on September 5, Ireland's Garda National Cyber Crime Bureau (GCCB) seized Conti's domains after its attack on the Irish HSE.[3] International sanctions soon followed. On September 21, 2021, the U.S. sanctioned SUEX, a cryptocurrency exchange implicated in laundering nearly $13 million for Conti and other ransomware operators.[4] Subsequently, on April 5, 2022, the German Federal Criminal Police Office dismantled Hydra Market, a darknet marketplace used by Conti to launder ransomware proceeds, confiscating 543 BTC worth approximately $23 million.[5] The U.S. then sanctioned Blender.io, a Bitcoin mixer, in part for laundering Conti proceeds.[6]

In addition to arrests and sanctions, government agencies worked to publicly expose Conti's tactics, techniques and procedures. On May 20, 2021, the FBI encouraged individuals to share intelligence on Conti's activities, increasing scrutiny.[7] A cybersecurity advisory by CISA, the FBI, and the NSA on September 22, 2021, detailed the threat posed by Conti and provided mitigation measures for organizations to protect themselves.[8] Similarly, Australian authorities sounded the alarm on December 10, 2021, highlighting a series of recent Conti attacks targeting multiple Australian organizations.[9] This increased public awareness sought to help organizations improve their defenses against Conti's evolving TTPs.

Recognizing the importance of targeting Conti's leadership, the U.S. State Department announced a $15 million reward on May 6, 2022, for information leading to the identification and capture of Conti leaders.[10] On August 11, 2022, a separate $10 million reward targeted five high-ranking Conti members, with their identities revealed publicly. Coordinated efforts between the U.S. and U.K. governments culminated in the February 9, 2023, sanctions against seven Conti and Trickbot members.[11] Eleven additional members were sanctioned on September 7, 2023, effectively outlawing any financial transactions involving these individuals.[12]

This final chapter expands the discussion from specific actions against Conti to a broader examination of how governments are tackling ransomware, highlighting current efforts and areas needing more focus. A three-pronged approach—disrupting ransomware groups' modus operandi, organizational structures, and branding—is essential for effectively combating these threats.[13] Governments have implemented various countermeasures to make it harder for ransomware groups to implement their operational playbook, such as issuing alerts, dismantling infrastructure, distributing decryption keys, and imposing sanctions on cryptocurrency exchanges.[14] They have also targeted internal hierarchies and affiliate networks through undercover operations, monitoring dark web forums, and public indictments. However, efforts to disrupt ransomware groups' branding and reputation have been insufficient. Current measures include doxing, taking down leak sites, and publishing reports on their failures, but more should be done. Governments should promote the establishment of a code of ethics for ransomware reporting, alongside a training program for journalists and other experts reporting on ransomware. This approach would help ensure more responsible coverage of ransomware activity, which in turn could reduce the undue glorification of these criminal groups and decrease their leverage over victims.

CONCLUSION

Countermeasures against Ransomware Modus Operandi

The initial set of approaches to counter ransomware focuses on disrupting the modus operandi of these groups by interfering with the specific tactics, techniques, and procedures (TTPs) they employ. In chapter two, I provided an overview of the stages of a ransomware attack. This approach targets each stage of a ransomware group's operational playbook to diminish their effectiveness and profitability. Table 6 outlines the various stages and the corresponding countering activities.

To tackle the early operational elements of ransomware, governments often issue warnings to alert organizations about new ransomware campaigns. This is typically based on intelligence shared between public and private sectors. An example is StopRansomware.gov, launched in 2021 by the U.S. government to help both public and private organizations defend against increasing ransomware attacks.[15] This whole-of-government initiative centralizes ransomware resources and alerts, guiding organizations to better understand ransomware threats, mitigate risks, and respond effectively to incidents. Since its launch, agencies like CISA and the FBI have issued numerous alerts, advisories, and updates on various ransomware strains and groups—including Conti, as well as BlackMatter, Play, and Scattered Spider—providing important information like indicators of compromise (IoCs), TTPs, and targeting activities.[16]

More proactive government measures have included the takedown of illicit online marketplaces that facilitate the access brokerage market. As briefly discussed, a prime example was the takedown of the Hydra market in April 2022. Operating since 2015, Hydra was the most prominent dark web marketplace in Russia and the largest globally, offering a platform for RaaS and other cybercriminal activities.[17] The US Treasury imposed sanctions on Hydra for facilitating cyber-enabled activities that threatened the US's security, policy, economic health, or financial stability. On the same day, German federal police seized Hydra's servers in Germany and confiscated 543 Bitcoins worth approximately $25 million at the time.[18] Similar disruptions were executed against Genesis Market in April 2023 and Breach Forum Domains in June 2023.[19]

These proactive measures also involve disrupting various botnet infrastructures, which are crucial for spreading ransomware.[20] A botnet takedown associated with Ryuk and later Conti involved Trickbot in October 2020.[21] As detailed in chapter four, Trickbot was initially released in 2016 to steal banking credentials. Over time, its capabilities expanded to collect Outlook credentials and other sensitive information from compromised machines, making it a favored tool for ransomware groups like Ryuk to deploy their payloads. By 2018, Trickbot had become one of the foremost cybersecurity threats and was later recognized as a significant risk to the integrity of the 2020 U.S. Presidential election process.[22] In early October 2020, Brian Krebs reported that an unknown entity was attempting to disrupt Trickbot by distributing fake configuration files to infected machines, redirecting them to a decoy control server.[23] The entity behind these efforts remained a mystery until mid-October when *The Washington Post* revealed that the U.S. Cyber Command had orchestrated the disruptions as part of a broader strategy of "persistent engagement" with cyber adversaries to protect the upcoming elections.[24] Concurrently, Microsoft undertook legal and technical actions against the botnet.[25]

In an attempt to disrupt ransomware groups' efforts to maintain persistence, establish command and control, escalate privileges, identify data for theft, and inhibit recovery—stages 4–6 of the ransomware playbook—governments have employed a mix of proactive measures and best practice promotions within organizations. One example of such proactive measure was the seizure of several Conti domains by the Irish government, which I detailed earlier in this chapter. The GCCB reported that the seizure directly prevented further ransomware incidents globally, with around 750 attempts to connect to these domains post-seizure, each potentially thwarting a ransomware deployment.

To assist organizations with decryption, governments have also at times succeeded in obtaining universal decryption keys for ransomware, which are discreetly shared with the victims.[26] For instance, in October 2022, Dutch police ingeniously tricked the

DeadBolt group into handing over decryption keys without any ransom being paid.[27] DeadBolt typically targeted smaller businesses and individuals, demanding relatively modest ransoms, but the high frequency of attacks—approximately 5000 known cases within a year—made their operations lucrative.[28] A critical vulnerability in their operations was their automatic decryption key delivery system upon payment. Dutch investigators exploited a flaw where DeadBolt's system sometimes released the decryption key before the victim's payment was confirmed in the blockchain, which usually takes about 10 minutes. The Dutch police developed a script that would initiate a payment, receive the decryption key, and then automatically cancel the payment before it was confirmed in the blockchain. This process not only helped victims retrieve their data but also prevented "hundreds of thousands of dollars" from flowing into DeadBolt's wallets before the group detected the anomaly.[29]

In another case, before taking down the Hive ransomware group in early January, the FBI infiltrated their servers for about seven months. During this time, the FBI created over 300 decryption keys and discreetly distributed them to victims, enabling them to unlock their systems without paying a ransom.[30] Additionally, the FBI supplied another 1,000 keys to individuals previously targeted by the group, helping them to retrieve some of their lost data. It saved victims from paying $130 million in demanded ransoms.

Governments occasionally face a challenging decision: they may possess decryption keys or know about an encryption flaw but choose not to immediately disclose this information publicly. This silence is intended to prevent tipping off the ransomware group about their access. For instance, in September 2021, Ellen Nakashima and Rachel Lerman of *The Washington Post* revealed that the FBI had withheld assistance for nearly three weeks following the major ransomware attack against Kaseya.[31] This delay was due to the FBI's plans to disrupt REvil, the group responsible for the attack.[32]

Governments typically advise against paying ransoms when it reaches the stage where criminals have issued a ransom demand.[33]

For instance, the Swiss National Cyber Security Center (NCSC) advises "not paying a ransom". As they elaborate, "once the ransom has been paid, there is no guarantee that the criminals will not publish the data anyway, or otherwise try to profit from it. Moreover, every successful ransom attempt motivates the attackers to continue, finances the further development of attacks and encourages their spread".[34] One of the most controversial and widely discussed governmental policies is the increasing call for laws that do not merely advise against, but outright prohibit ransom payments. Former UK NCSC head Ciaran Martin has argued for the effectiveness of such a ban: "we have to find a way of making a ransomware payments ban work".[35] In 2023, Emisoft, a software company known for its decryption services, argued that "We believe that the only solution to the ransomware crisis—which is as bad as it has ever been—is to completely ban the payment of ransoms".[36]

While this approach directly targets the ransomware payment model, it also presents potential adverse consequences. Implementing such a policy without exceptions is challenging, especially for critical sectors like healthcare, where decryption might be the only option to save lives.[37] This likely leads to ransomware groups targeting the very sectors where payments are still permitted, which are often the most vulnerable and critical.[38]

Lastly, in addressing the issue of ransomware payouts, authorities have targeted cryptocurrency exchanges, which are often used for the illicit transfer of funds. An early example is the May 2013 takedown of Liberty Reserve. Based in Costa Rica, Liberty Reserve allowed users to open accounts and transfer money with minimal identification requirements—a name, date of birth, and an email address.[39] Its relative anonymity made it a favorite among criminals, facilitating the transfer of approximately $6 billion in illicit proceeds over seven years.[40] The founder, Arthur Budovsky Belanchuk, was arrested in Spain on charges of money laundering, and shortly thereafter, the Liberty Reserve website was taken offline, displaying a message that it had been seized by U.S. authorities.[41] In 2019, the data collected from Liberty Reserve also contributed to the arrest and conviction of a member of the Reveton ransomware group.[42]

Another significant case occurred in March 2023 with the take-down of Chip Mixer. Described as a "darknet cryptocurrency mixing service", Chip Mixer was responsible for laundering more than $3 billion worth of cryptocurrency over about six years, sourced from activities including ransomware attacks and darknet market transactions.[43] Chip Mixer's services were popular among ransomware and other cybercriminal groups for their ability to obscure blockchain trails, making tracking by law enforcement challenging, if not impossible. On March 15, German and U.S. authorities, with support from Europol and agencies in Belgium, Switzerland, and Poland, announced the seizure of four Chip Mixer servers located in Germany, approximately $46.5 million in Bitcoin, and seven terabytes of data.[44] Additionally, Minh Quốc Nguyễn, the Vietnamese operator of the platform, was charged with money laundering by US authorities in Philadelphia.[45]

Table 6: List of key government countermeasures per stage of ransomware operation

Stage	Countermeasures
1. Reconnaissance	• Threat intelligence sharing between private and public sectors to preemptively identify and counter-act reconnaissance efforts by cybercriminals. • Proactive alerts about new ransomware campaigns.
2. Network Compromise	• Cyber hygiene and awareness campaigns. • Disruption or takedown access brokering or botnet infrastructure.
3. Situational Awareness and Privilege Escalation	• Promotion of regular security audits.
4. Persistence and C2	• Disruption or takedown of C2 infrastructure.

5. Identification Directories	• Promotion of cybersecurity frameworks that include network segmentation.
6. Exfiltration of Data	• Cybersecurity frameworks that include backups.
7. Encryption	• Distribution of obtained decryption tools through controlled channels.
8. Negotiations	• Establishment of specialized government units that provide expert advice to (certain) ransomware victims during negotiations. • Provision of legal guidance for negotiations.
9. Payment or Publication	• Takedowns or blocking access to websites that host stolen data as part of extortion schemes.
10. Money Transfers	• Introduction and enforcement cryptocurrency exchange regulation.
11. Cash Out	• Seizure or freezing of cryptocurrency assets.

Countermeasures against Organizational Structure

Disrupting the organizational structure of ransomware groups focuses on dismantling their internal hierarchies and affiliate networks. This approach aims to destabilize the ransomware group's organizational structure and prevent them from coordinating effective ransomware attacks.

Classical police work and criminal prosecution, including investigating, charging and arresting perpetrators, was one of the earliest and still persistent actions governments take against ransomware. For example, the man behind the first known ransomware incident, Joseph L. Popp, was tracked down, arrested and prosecuted after having distributed around 20,000 malicious floppy disks in 1989.

However, since these early days, the growth of and access to and through the internet has dramatically changed the nature of ransomware and the criminal prosecution of its perpetrators. Additional investigative skills, such as digital forensics, became a necessity. However, even if investigators gather, analyse, and follow-up on digital clues, being able to reach the perpetrators in the physical world and completely stop the ransomware has become a very difficult, if not impossible, endeavor. Even just arresting the responsible individuals has become incredibly hard. One rather rare example of successful arrest is certainly the 2013 apprehension of eleven individuals who were behind a surge of the prolific "Police Ransomware" in Spain and across Europe in from 2011 onwards. This ransomware usually claimed to be an official police agency, typically referencing and including official local police and their emblems on the ransom note. In this case, the police were able to apprehend (at least one of) the actual developers of the ransomware; with the emergence of the RaaS model, this is rare and, most often, it is the lower ranked operators or affiliates executing the attacks or laundering money who are arrested—as we saw with REvil.[46]

Although arrests remain a rare occurrence, governments, especially the U.S., still often charge core individuals with the crimes they have committed and sometimes even put out financial rewards for any helpful information regarding these individuals. Indeed, this is exactly what happened in the case of Conti in 2022, when the U.S. identified five of its leading members and put out a $10 million reward for any information on them.[47]

Finally, what may very well happen—though there are fewer publicly known examples of this—is government agencies trying to sow distrust within ransomware groups. This can be achieved by leaking information about potential informants, spreading misinformation about alliances, or manipulating communications within the group. The goal is to make members doubt each other's loyalty and intentions, which can hamper their coordination and effectiveness.

Countermeasures against Branding and Reputation

As I emphasized throughout this book, branding and reputation are crucial for ransomware groups in overcoming the *Ransomware Trust Paradox*. Ransomware groups engage in inherently deceptive and extortionate activities, such as breaking into systems, stealing data, and encrypting vital information. Despite these actions, they must also convince their victims of their reliability. This trust is crucial not only for persuading victims not to release stolen data, but also for ensuring that payments will lead to the decryption of affected systems. Thus, branding and reputation-building are central, not peripheral, to the success of these groups.

However, governments and organizations often overlook efforts to undermine this aspect of ransomware groups. When the International Counter Ransomware Initiative (CRI) was launched in 2021 by the White House and international partners, it prioritized enhancing network resilience, disrupting ransomware operations through law enforcement collaboration, and countering the financial mechanisms that sustain ransomware profitability.[48] At the second CRI Summit, members reaffirmed their commitment to these goals and established the International Counter Ransomware Task Force (ICRTF) to coordinate and disrupt ransomware activities on an operational level.[49] By the third meeting in 2023, the CRI expanded its scope, incorporating capacity-building efforts and private sector collaboration into all aspects of its strategy.[50] Despite these advances, the CRI has not yet addressed the Ransomware Trust Paradox and the crucial role that branding and reputation play in the success of ransomware groups.

Similarly, the Ransomware Task Force (RTF) a commendable multi-stakeholder effort by the Institute for Security and Technology (IST), includes participants from government, industry, and civil society.[51] On May 2021, the RTF released a thorough report with 48 key recommendations aimed primarily at governments and the private sector. This report has significantly improved efforts to disrupt ransomware operations, enhance information sharing, and strengthen strategies for mitigating and recovering from attacks.[52] Yet, like the CRI, the RTF's report offers no policy recommendations to undermine the reputation or credibility of ransomware groups.[53]

We have only seen sporadic attempts to undercut ransomware group's credibility and reliability.[54] For instance, government disclosures have highlighted cases where ransomware groups failed to decrypt data after receiving payments or targeted victims again, despite promises to the contrary. Similarly, after the UK National Crime Agency (NCA) and its international partners disrupted Lockbit, officials have stressed on numerous occasions that the ransomware group does not fully delete data even after receiving a payment, contradicting their previous assurances.[55] This revelation has also made it harder for Lockbit to make a comeback; even if it manages to bring some infrastructure back up, it still carries a stigma that undermines its operations. Negotiators have also begun using this knowledge in subsequent cases with other ransomware groups, leveraging it to secure lower ransom demands by challenging the credibility of attackers' promises based on this precedent.[56]

Additionally, there have been instances where connections between different ransomware brands have been revealed, showing that some groups target the same victims under different guises.[57] For example, in the chapter on Conti's business expansion, I detail how the group's ties with Diavol were exposed, ties that Stern was determined to keep hidden from the public.

However, these efforts should go much further. Over the past few years, the lifecycle of ransomware brands has been shorter. New ransomware brands seem to come and go more quickly. This volatility can be attributed to increased competition within the sector and more suppliers to outsource operations and tooling, leading to a less monopolized market and, consequently, faster turnover of ransomware brands. Additionally, intensified government efforts to combat ransomware as discussed in this chapter have influenced these dynamics. However, it has also become notably easier for a ransomware newcomer to quickly build brand equity, thanks to the burgeoning industry dedicated to ransomware reporting. Previously, ransomware groups had to make substantial efforts to garner public attention, in a time when cybersecurity coverage was not the multi-billion-dollar industry it is today, relying on a small cadre of security experts for exposure. Now, the scenario has transformed: any release by a ransomware group on a leak site gains immediate traction across social media platforms and news websites, as well as expert company analyses.

The race to cover new entrants in the cybercriminal ecosystem introduces a unique and complex challenge for the public dissemination of information on ransomware, unlike the situation with APTs. This difference arises from the contrasting preferences of these groups; APTs aim to operate under the radar, whereas ransom war groups often seek out and benefit from the visibility provided by cybersecurity reporting. Consequently, the cybersecurity community's eagerness to report plays a critical role in shaping the ransomware ecosystem. Public reporting speeds up the rebranding efforts of these groups but also aids in the development of their reputations, and significantly affects their positioning within the constantly shifting cybersecurity environment and relationships with victims.

Indeed, Conti's notoriety—known for being both efficient and ruthless—was largely shaped by media narratives and research analyses. Between 2020 and 2021, over a thousand articles on various aspects of Conti's operations were published. This extensive coverage enhances Conti's brand and even featured in their ransom note; "if you don't who we are—just 'google it'". Similarly, in one of the blog posts, Vice Society thanks a specific journalist for an article in which it was part of a 'Top 5' of ransomware and malware groups in 2022.[58] This means journalists and cybersecurity researchers need to be cautious when reporting on ransomware attacks to avoid inadvertently promoting the groups behind these attacks.[59]

One pertinent example is the realm of disinformation, where media coverage can inadvertently reinforce the very dynamics it seeks to dismantle.[60] To address this challenge, UNESCO developed a handbook for journalism training on fake news, prompted by growing international concerns over a 'disinformation war' targeting journalists and the broader media landscape. This handbook highlights seven core principles—fairness, independence, accuracy, contextuality, transparency, protection of confidential sources, and perspicacity - that collectively build trust, credibility, and public confidence.[61] Moreover, a report by Whitney Phillips from Data & Society encourages journalists to assess whether a story has reached a tipping point, offers a public health takeaway, provides a political or social action point, and whether the risk of entrenching or rewarding the falsehood outweighs the benefits of debunking it.[62] If these criteria are not met, the story might best be left unreported at that time.

Perhaps even more instructive is the field of counterterrorism. Terrorists seek to instill widespread fear, and while accurate reporting on terrorist incidents is essential, it can paradoxically help achieve that aim. Consequently, a range of guidelines and training workshops have been developed to help journalists navigate this sensitive reporting terrain, ensuring that coverage does not inadvertently advance terrorists' objectives.[63]

There are at least five critical elements that should form the foundation of any code of ethics for ransomware reporting. First, accuracy must be prioritized over speed. For example, the rise of automated bots that scrape and publish information from leak sites without verification is problematic. These bots can quickly spread misinformation, which is often unverified and potentially misleading, released by ransomware groups aware of the attention these sites garner. For instance, Andy Greenberg at *WIRED* exemplified responsible journalism when he stated, "We at *WIRED* held off on reporting on the second ransomware gang, RansomHub, threatening to leak a trove of stolen data from Change Healthcare until the hackers provided evidence of their claims".[64] Second, journalists must use language cautiously when describing ransomware groups. It is important to avoid terms that might inadvertently glorify these groups or enhance their fearsome image. Such portrayals can play into the hands of ransomware operators by bolstering their intended reputation as formidable and fearless. Third, the reporting should aim to balance its focus and avoid overemphasizing successful attacks by ransomware groups. It is beneficial to highlight instances where security measures or backup strategies have successfully mitigated the impact of an attack, or where organizations have managed to recover without complying with ransom demands.[65] This helps to provide a more balanced view of ransomware's actual threat level and the effectiveness of protective measures. Fourth, like any cyber incident, careful consideration should be given to the amount of technical detail shared about security vulnerabilities. Information about unpatched vulnerabilities should only be published if it is essential for public safety and awareness, ensuring that it does not provide cybercriminals with exploitable information.

Cherry Picking and International Partnerships

While this chapter presents prominent initiatives from around the world, it does not fully represent the global effort required. As I noted in the chapter on pioneering, Russia has shown a lack of initiative in countering ransomware originating from its territory.[66] Yet, many governments in the West and other parts of the world also still need to significantly enhance their anti-ransomware measures and commitment. In other words, my cherry picking of examples should not obscure the fact that many governments still do not have a comprehensive national strategy to address ransomware.

There is also a pressing need for greater unification of government responses. Cyberspace transcends traditional nation-state borders, making it easy for cybercriminals to operate from one country while targeting victims in others. Recognizing this, governments have stressed the importance of international cooperation from the outset as a cornerstone of major actions against ransomware. Established international law enforcement bodies like Europol and Interpol have played key roles in coordinating comprehensive actions over the years.[67] Indeed, almost all significant operations against ransomware have involved multiple agencies from various governments. Beyond these institutions, there has been a rise in direct cooperation and information exchange specifically focusing on ransomware, exemplified by the establishment of the International Counter Ransomware Initiative and its Counter Ransomware Task Force, which includes a growing number of countries. It is these efforts that need both *continuation* and *expansion* if we are to effectively combat the global ransomware threat.[68]

CONCLUSION

The Ransomware Trust Paradox offers a unique opportunity for this expanded intervention. We must move beyond traditional tactics and consider the nuanced role of trust in the ransomware ecosystem. By adopting an approach that not only targets the operational capabilities of these groups, but also diminishes their ability to project trustworthiness and credibility, we can undermine the very foundations of ransomware groups' profitability. As we refine our counter-ransomware efforts, it is essential to remain vigilant against inadvertently strengthening these groups through media coverage, which underscores the need for a carefully crafted code of ethics.

ACKNOWLEDGEMENTS

After finishing my book *No Shortcuts: Why States Struggle to Develop a Military Cyber-Force* at the end of 2021, I faced a dilemma. I was keen to keep exploring military and intelligence operations in cyberspace, especially with new developments like the United Kingdom's National Cyber Force creation. I was also curious about how militaries use data analytics to make better decisions and improve their operations. However, the increasing threat of ransomware kept pulling my attention. It was clear from all the news about how healthcare institutions and other organizations were suffering that ransomware was not just a business problem—it was a national security crisis that was getting worse. It deserved more attention, not just from incident responders but also from other fields.

As I started working on this book in early 2022, I noticed something surprising: hardly any academic articles on ransomware had been published in Political Science or International Relations, my PhD field. This was still true when I handed my manuscript to the publisher in mid-2024. Given how much ransomware threatens our security, we need to dig deep into how these criminal groups operate and affect society—and turn this into a field of study.

I am thankful to a remarkable group of people whose support was pivotal in bringing this book to life. A heartfelt thank you to Michael Dwyer at Hurst for embracing the manuscript with enthusiasm and his dedication in bringing this project to fruition. The research environment at ETH Center for Security Studies helped me focus on writing, complemented by great research assistance

ACKNOWLEDGEMENTS

from Oliver Roos and Janina Inauen. I appreciate Alex Bollfrass for his creative input in titling the book. My thanks also extend to Eugenio Benincasa, Dan Black, Jamie Collier, Myriam Dunn Cavelty, Florian Egloff, Jen Ellis, Juan Andres Guerrero Saade, Jamie MacColl, Jiro Minier, Jasmin Stadler, James Shires, Will Lyne, and Nicolas Zahn for their insightful comments on the manuscript drafts, and to everyone else who engaged in discussions and provided feedback that helped refine my ideas. Most importantly, I want to thank my wife, Diana, for her incredible patience and support. I am looking forward to spending more weekends away from my desk on hikes, rides and swims.

NOTES

INTRODUCTION

1. Ransomware is an amalgamation of ransom and malware. According to United States US Cybersecurity & Infrastructure Security Agency (CISA), "Ransomware is an ever-evolving form of malware designed to encrypt files on a device, rendering any files and the systems that rely on them unusable. Malicious actors then demand ransom in exchange for decryption. Ransomware actors often target and threaten to sell or leak exfiltrated data or authentication information if the ransom is not paid. In recent years, ransomware incidents have become increasingly prevalent among the Nation's state, local, tribal, and territorial (SLTT) government entities and critical infrastructure organizations". CISA, "Ransomware 101", Stopransomware.org, https://www.cisa.gov/stopransomware/ransomware-101. For an etymology of ransomware see: Allan Liska, "The Etymology of Ransomware", *Ransomware*, June 11, 2023, https://ransomwaresommelier.com/p/the-etymology-of-ransomware.
2. El País, "Costa Rica Rechaza Pagar a Ciberdelincuentes por Secuestro de Datos", April 20, 2022, https://web.archive.org/web/20220420211937/https://www.elpais.cr/2022/04/20/costa-rica-rechaza-pagar-a-ciberdelincuentes-por-secuestro-de-datos/
3. VenariX, X, April 18, 2022, https://x.com/_venarix_/status/1516113184633110529
4. Vitali Kremez, Yelisey Boguslavskiy, and Marley Smith, "Anatomy of Attack: Truth behind the Costa Rica Government Ransomware 5-Day Intrusion", Internet Archive: WayBackMachine, November 26, 2022, https://web.archive.org/web/20221126092151/https://www.advintel.io/post/anatomy-of-attack-truth-behind-the-costa-rica-government-ransomware-5-day-intrusion
5. Ionut Ilascu, "How Conti Ransomware Hacked and Encrypted the Costa Rican Government", *Bleeping Computer*, July 21, 2022, https://www.bleepingcomputer.com/news/security/how-conti-ransomware-hacked-and-encrypted-the-costa-rican-government/.

6. "Central Bank Intensifies Surveillance of its Systems Due to &#82 16;Hacks’ in Costa Rica by Conti Group". *CE Noticias Financieras English*, April 19, 2022, https://advance.lexis.com/api/document?collection=news& id=urn:contentItem:658D-3021-JCG7–800G-00000–00&context=1516831.

7. Hereafter Twitter is only referred to as "X". "CCSS Twitter Account was Hacked by Cyber Hackers", *CE Noticias Financieras English*, April 19, 2022, https:// advance.lexis.com/api/document?collection=news&id=urn:contentItem:6 58D-3011-JCG7–8454–00000–00&context=1516831.

8. Rico, "Conti Cyber Attacks in Costa Rica Is Now against Eight Public Institutions", *Q COSTA RICA*, April 27, 2022, https://qcostarica.com/ conti-cyber-attacks-in-costa-rica-is-now-against-eight-public-institutions/.

9. Jonathan Greig, "Conti Ransomware Cripples Systems of Electricity Manager in Costa Rican Town", *The Record*, April 25, 2022, https://therecord.media/ conti-ransomware-cripples-systems-of-electricity-manager-in-costa-rican-town

10. The Ministry of Finance was the one hit hardest. Rico, "President Chaves: 'We Are at War and That Is Not an Exaggeration'", *Q COSTA RICA*, May 16, 2022, https://qcostarica.com/president-chaves-we-are-at-war-and-that-is-not-an-exaggeration/

11. Javier Córdoba, "Costa Rica Declares Emergency in Ongoing Cyber Attack", *AP News*, May 12, 2022, https://apnews.com/article/covid-technology-health-caribbean-costa-rica-949b141c5b5e288214f80d2cb6753bdb

12. The term has earlier references to significant ransomware incidents. For an overview, see Will Thomas' talk at Sleuthcon: Will Thomas, "Xakep, Repa, Probiv, Spy", Sleuthon, May 24, 2024. Also see: Nissim Ben Saadon, "Ransom-War Escalation: The New Frontline in Cyber Warfare", *Cyber Defence Magazine*, February 22, 2024, https://www.cyberdefensemagazine.com/ransom-war-escalation-the-new-frontline-in-cyber-warfare/?utm_source=dlvr.it&utm_medium=twitter; Robin Pomeroy, host. "Ransomware and 'Ransom-War': Why We All Need to Be Ready for Cyberattacks", Radio Davos, *World Economic Forum*, April 1, 2022, https://www.weforum.org/podcasts/radio-davos/episodes/cybersecurity/; Natto Team, "Ransom-War: Russian Extortion Operations as Hybrid Warfare, Part One", *Natto Thoughts*, May 1, 2024, https://nattothoughts.substack.com/p/ransom-war-russian-extortion-operations; James McQuiggan. "Ransomware, Ransom-War and Ran-some-where: What We Can Learn When the Hackers Get Hacked", *KnowBe4*, 2022, https://info.knowbe4. com/ransomware-ransom-war

13. Rico, "Government Rules out Paying Extorsion to Cybercriminals", *Q COSTA RICA*, April 21, 2022, https://qcostarica.com/government-rules-out-paying-extorsion-to-cybercriminals/

14. Ned Price, "Reward Offers for Information to Bring Conti Ransomware Variant Co-Conspirators to Justice", United States Department of State, May 6, 2022, https://www.state.gov/reward-offers-for-information-to-bring-conti-ransomware-variant-co-conspirators-to-justice/

15. VenariX, X, April 25, 2022, https://x.com/_venarix_/status/151870055 0543581186

16. The first reference to Biden was published on May 8 and was possibly a reaction to the U.S. government's May 6 reward announcement. See Price, "Reward Offers for Information to Bring Conti Ransomware Variant Co-Conspirators to Justice"; VenariX, X, May 8, 2022, https://x.com/_venarix_/status/152 3410318206009344

17. Ministerio de Hacienda de Costa Rica, X, September 10, 2023, https://x.com/ HaciendaCR/status/1516190939114803203

18. Rico, "Treasury Computer Systems Hacked!", *Q COSTA RICA*, April 19, 2022, https://qcostarica.com/treasury-computer-systems-hacked/

19. CGR, "Informe de Auditoria sobre la seguridad de la informacion de los sistemas del ministerio de hacienda", April 28, 2023, https://cgrfiles.cgr.go.cr/ publico/docs_cgr/2023/SIGYD_D/SIGYD_D_2023001310.pdf

20. Daniela Cerdas E., "Informe Destapa Crisis en el MEP Por Ataque de Conti", *La Nación*, May 6, 2022, https://www.nacion.com/el-pais/educacion/hackeo-paraliza-pagos-de-salarios-nombramientos-e/GDKBBW2QRNAU 3IRW5EGCPVUG6M/story/

21. DELFINO, "Hackeo de Conti ha Afectado Pagos de 12 Mil Docentes, MEP Volverá a Planilla Manual Para Resolver Crisis", May 17, 2022, https://delfino.cr/2022/05/hackeo-de-conti-ha-afectado-pagos-de-12-mil-docentes-mep-volvera-a-planilla-manual-para-resolver-crisis

22. Karina Porras Díaz, "Grupo de Educadores se Manifestará Este Martes Frente al MEP Por Atrasos en el Pago de Salarios", *Monumental*, May 16, 2022, https:// www.monumental.co.cr/2022/05/16/grupo-de-educadores-se-manifestara-este-martes-frente-al-mep-por-atrasos-en-el-pago-de-salarios/

23. "Costa Rica Registers Cyberattack; State Digital Platforms Suspended", *CE Noticias Financieras English*, April 20, 2022, https://advance.lexis.com/api/doc ument?collection=news&id=urn:contentItem:658M-21V1-DYY9–02B0– 00000–00&context=1516831

24. "'Hacking' at the Treasury Department Forces Creation of a 'Manual' Process to Reactivate Imports", *CE Noticias Financieras English*, April 20, 2022, https://advance.lexis.com/api/document?collection =news&id=urn:contentItem:658M-21T1-DYY9–00W0–00000– 00&context=1516831

25. DELFINO, "Cámaras Empresariales Piden Declaratoria de Emergencia Nacional Por Situación en Aduanas", May 3, 2022, https://delfino.cr/2022/05/cama-ras-empresariales-piden-declaratoria-de-emergencia-nacional-por-situacion-en-aduanas

26. "'We Will Continue Attacking Ministries Until Your Government Pays Us':·Conti's Threat to Costa Rica", *CE Noticias Financieras English*, April 19, 2022, https://advance.lexis.com/api/document?collection=news&id=u rn:contentItem:658M-21V1-DYY9–00XW-00000–00&context=1516831;

Ministerio de Hacienda de Costa Rica, X, April 26, 2022, https://x.com/HaciendaCR/status/1518937945738207233

27. VenariX, X, April 25, 2022, https://x.com/_venarix_/status/1518605134300917760; VenariX, X, April 27. 2022, https://x.com/_venarix_/status/1519410720802287616

28. After the Estonian government relocated a six-foot-tall bronze statue, which commemorated the Soviet victory over Nazi Germany, from the city center to a cemetery on the outskirts of Tallinn, several political, financial, and media websites, along with other electronic services, were attacked by predominantly Russian-based groups and individuals using well-known methods. These Distributed Denial of Service (DDoS) attacks unfolded over a period of 22 days. While *La Nación* drew parallels, the DDoS attacks against Estonia were significantly less severe. "Editorial: Urge Invertir en Seguridad Informática", *La Nación*, April 26, 2022, https://www.nacion.com/opinion/editorial/editorial-urge-invertir-en-seguridad-informatica/A2PMA6LPU5EEVAC2F5U3PSHKNY/story/; Rain Ottis, "Analysis of the 2007 Cyber Attacks Against Estonia from the Information Warfare Perspective", In Proceedings of the 7th European Conference on Information Warfare, Reading (MA: Academic Publishing Limited, 2008), 163.

29. VenariX, X, April 27, 2022, https://x.com/_venarix_/status/1519410720802287616;VenariX, X, April 28, 2022, https://x.com/_venarix_/status/1519760155176488960

30. VenariX, X, May 14, 2022, https://x.com/_venarix_/status/1525457045826195456

31. VenariX, X, May 14, 2022, https://x.com/_venarix_/status/1525539124115038210

32. VenariX, X, May 21, 2022, https://x.com/_venarix_/status/1527805232515887114

33. Yelisey Bogusalvskiy, and Vitali Kremez, "DisCONTInued: The End of Conti's Brand Marks New Chapter For Cybercrime Landscape", Internet Archive: WayBackMachine, October 26, 2022, https://web.archive.org/web/20221026025639/https://www.advintel.io/post/discontinued-the-end-of-conti-s-brand-marks-new-chapter-for-cybercrime-landscape; for follow-on activity from the US State Department, also see: Martin Matishak, "U.S. Convenes 30 Countries on Ransomware Threat—without Russia or China", October 13, 2021, https://therecord.media/u-s-convenes-30-countries-on-ransomware-threat-without-russia-or-china

34. Alexander Martin, "Ransomware Incidents Now Make up Majority of British Government's Crisis Management 'Cobra' Meetings", *The Record*, November 18, 2022, https://therecord.media/ransomware-incidents-now-make-up-majority-of-british-governments-crisis-management-cobra-meetings

35. National Cyber Security Centre, *Annual Review 2022: Making the UK the Safest Place to Live and Work Online*, 2022, https://www.ncsc.gov.uk/files/NCSC-Annual-Review-2022.pdf

36. Luke Irwin, "South Staffordshire Water Targeted by Cyber Attack", *IT Governance UK Blog*, August 16, 2022, https://www.itgovernance.co.uk/blog/south-staffordshire-water-targeted-by-cyber-attack

37. "Internet Organised Crime Threat Assessment (IOCTA) 2020", Europol, December 7, 2021, https://www.europol.europa.eu/publications-events/main-reports/internet-organised-crime-threat-assessment-iocta-2020. On the various harms from ransomware attacks also see: Nandita Pattnaik, Jason R.C. Nurse, Sarah Turner, Gareth Mott, Jamie MacColl, Pia Huesch, and James Sullivan, "It's more than just money: The real-world harms from ransomware attacks", 17th International Symposium on Human Aspects of Information Security & Assurance (HAISA 2023), 2307.02855 (arxiv.org)

38. Alexander Martin, "Ransomware Gang Posts Breast Cancer Patients' Clinical Photographs", *The Record*, March 6, 2023, https://therecord.media/ransomware-lehigh-valley-alphv-black-cat; Marianne Kolbasuk McGee, "Breast Cancer Patients Sue Over Breached Exam Photos, Data", *Bank Info Security*, March 14, 2023, https://www.bankinfosecurity.com/breast-cancer-patients-sue-over-breached-exam-photos-data-a-21431

39. AP News, "Change Healthcare cyberattack was due to a lack of multifactor authentication, UnitedHealth CEO says", May 1, 2024, https://apnews.com/article/change-healthcare-cyberattack-unitedhealth-senate-9e2fff-70ce4f93566043210bdd347a1f

40. It remarkably led to the implosion of the BlackCat/ALPHV, the ransomware group that was behind the extortion. The hacker responsible for facilitating BlackCat's initial access into Change Healthcare's network has accused the criminal group of withholding their portion of the ransom payment. Additionally, this individual asserts that BlackCat retained the sensitive data that Change allegedly compensated the group to destroy. This disclosure has reportedly caused BlackCat to stop its operations. Brian Krebs, "BlackCat Ransomware Group Implodes After Apparent $22M Payment by Change Healthcare", Krebson Security, March 5, 2024, https://krebsonsecurity.com/2024/03/blackcat-ransomware-group-implodes-after-apparent-22m-ransom-payment-by-change-healthcare/

41. Chris Vallace and Joe Tidy, "Ransomware Attack hits Dozens of Romanian Hospitals", *BBC News*, February 13, 2024, https://www.bbc.com/news/technology-68288150

42. Claire McGlave, Hannah Neprash and Sayeh Nikpay, "Hacked to Pieces? The Effects of Ransomware Attacks on Hospitals and Patients", October 4, 2023, *SSRN*, https://ssrn.com/abstract=4579292 or http://dx.doi.org/10.2139/ssrn.4579292

43. This includes my own writing: e.g. Max Smeets, *No Shortcuts: Why States Struggle to Develop A Military Cyber-Force* (Oxford University Press, 2022); Robert Chesney and Max Smeets, eds. *Deter, Disrupt, or Deceive: Assessing Cyber Conflict as an Intelligence Contest* (Georgetown University Press, 2023); Richard Harknett

and Max Smeets, "Cyber Campaigns and Strategic Outcomes", *Journal of Strategic Studies*, 45, no. 4 (2020), https://doi.org/10.1080/01402390.2020.1732354

44. See, for example: Erik Gartzke, "The Myth of Cyberwar: Bringing War in Cyberspace Back Down to Earth", *International Security* 38, no. 2 (October 2013): 41–73, https://doi.org/10.1162/ISEC_a_00136; Adam P. Liff, "Cyberwar: A New 'Absolute Weapon'? The Proliferation of Cyberwarfare Capabilities and Interstate War", *Journal of Strategic Studies* 35, no. 3 (June 2012): 401–28, https://doi.org/10.1080/01402390.2012.663252. Also see: Timothy J. Junio, "How Probable Is Cyber War? Bringing IR Theory Back In to the Cyber Conflict Debate", *Journal of Strategic Studies* 36, no. 1 (February 2013): 125–33, https://doi.org/10.1080/01402390.2012.739561; Adam P. Liff, "The Proliferation of Cyberwarfare Capabilities and Interstate War, Redux: Liff Responds to Junio", *Journal of Strategic Studies* 36, no. 1 (February 2013): 134–38, https://doi.org/10.1080/01402390.2012.733312; Martin Libicki, *Cyberwar and Cyberdeterrence* (Santa Monica: RAND Corporation, 2009), https://www.rand.org/content/dam/rand/pubs/monographs/2009/RAND_MG877.pdf; also see: Martin C. Libicki, "Cyberspace Is Not a Warfighting Domain", *I/S A Journal of Law and Policy for the Information Society* 8, no. 2 (2012): 325–40, http://moritzlaw.osu.edu/students/groups/is/files/2012/02/4.Libicki.pdf

45. "NATO Agency to Participate in World's Largest Cyber Security Exercise Locked Shields", NATO Communications and Information Agency, April 14, 2022, https://www.ncia.nato.int/about-us/newsroom/nato-agency-to-participate-in-worlds-largest-cyber-security-exercise-locked-shields.html; Max Smeets, "The Role of Military Cyber Exercises: A Case Study of Locked Shields", edited by T. Jančárková, G. Visky, and I. Winther, 2022.

46. In chapter two I provide a more extensive discussion of the differences between APTs and ransomware groups. Black Hat, *Keynote: Black Hat at 25: Where Do We Go from Here?* Youtube, November 18, 2022, https://www.youtube.com/watch?v=doRZwCbbyNs

47. "National Cyber Threat Assessment 2023–2024", Canadian Centre for Cyber Security, October 28, 2022, https://www.cyber.gc.ca/en/guidance/national-cyber-threat-assessment-2023–2024

48. For a similar discussion on terrorism see: Jacob N. Shapiro, *The Terrorist's Dilemma: Managing Violent Covert Organizations* (Princeton University Press, 2013).

49. To be sure, individuals can still engage in ransomware activities and cause significant harm. An illustrative example occurred in October 2020 when the Finnish police revealed a breach involving the treatment records of tens of thousands of psychotherapy patients belonging to the private company Vastaamo, which owns twenty-five therapy centers in Finland. The perpetrator first contacted Vastaamo's staff, demanding a ransom of €450,000 in exchange for the patients' medical records. When Vastaamo refused to comply, the hacker

resorted to leaking the data online and blackmailing the patients directly, demanding €200 in Bitcoin to prevent the public release of their records. The country's Minister of Interior, Maria Karoliina Ohisalo, who is responsible for issues related to internal security, describes the hack as "a shocking act which hits all of us deep down". On October 28, 2022, the National Bureau of Investigation identified the suspect behind the breach as 25-year-old Aleksanteri Julius Kivimäki, who was subsequently arrested a couple of months later in France. While there will always be lone wolf attacks, it is the organized crime aspect of ransomware that poses a recurring threat to large companies and society as a whole, extorting millions of dollars repeatedly and implementing systems to cash out these payments. See: AFP, "'Shocking' Hack of Psychotherapy Records in Finland Affects Thousands", *The Guardian*, October 26, 2020, sec. World news, https://www.theguardian.com/world/2020/oct/26/tens-of-thousands-psychotherapy-records-hacked-in-finland

50. Some have put the estimate at 600. Yaara Shriebman, "Ransomware 2021—The Bad, The Bad & The Ugly", Cyberint, December 30, 2021, https://cyberint.com/blog/research/ransomware-2021-the-bad-the-bad-the-ugly/; others at 1000: "Conti Ransomware Gang Shutdown, Conti Ransomware Rebranding 2022", *CyberTalk*, May 20, 2022, https://www.cybertalk.org/2022/05/20/conti-ransomware-gang-shuts-down-rebranding-into-smaller-units/

51. Estimates for Conti vary greatly, with some projections reaching as high as 2.7 billion. See: Monica Pitrelli, "Leaked Documents Show Notorious Ransomware Group Has an HR Department, Performance Reviews and an 'Employee of the Month'", *CNBC*, April 13, 2022, https://www.cnbc.com/2022/04/14/conti-ransomware-leak-shows-group-operates-like-normal-tech-company.html

52. Price, "Reward Offers for Information".

1. FROM RANSOMWARE TO RANSOM WAR GROUPS

1. "AIDS 88 Summary: A Practical Synopsis of the IV International Conference, Stockholm, Sweden", NCJRS Virtual Library, 1988, https://www.ojp.gov/ncjrs/virtual-library/abstracts/aids-88-summary-practical-synopsis-iv-international-conference

2. Edward Wilding, "The Authoritative International Publication on Computer Virus Prevention, Recognition and Removal", *Virus Bulletin*, January 1992, https://www.virusbulletin.com/uploads/pdf/magazine/1992/199201.pdf

3. See Renee Dudley and Daniel Golden, *The Ransomware Hunting Team: A Band of Misfits' Improbable Crusade to Save the World from Cybercrime* (Farrar, Straus and Giroux, 2022); David Ferbrache, *A Pathology of Computer Viruses* (Springer Science & Business Media, 1992).

4. Aimee Bosman, "First Ransomware: Joseph L. Popp", LinkTek.com, February 19, 2024, https://linktek.com/malice-money-monkeys-and-a-madman-the-origin-of-ransomware/

5. Dudley and Golden, *The Ransomware Hunting Team*.

6. Edward Wilding, "The Authoritative International Publication on Computer Virus Prevention, Recognition and Removal", *Virus Bulletin*, March 1990, https://www.virusbulletin.com/uploads/pdf/magazine/1990/199003.pdf

7. Dudley and Golden, *The Ransomware Hunting Team*.

8. Ibid.

9. Edward, "The Authoritative International Publication".

10. While Popp was in a psychiatric hospital, he was overheard on the telephone bragging to someone that he had deceived the system to evade trial. The Southwark judge, however, did believe Popp's claim. See Dudley and Golden, *The Ransomware Hunting Team*.

11. For discussion on some of the general technological drivers for cybercrime and ransomware see: Matthew Ryan, *Ransomware Revolution: The Rise of a Prodigious Cyber Threat* (Springer: 2021).

12. "Sophos Cracks Archiveus Ransomware Code", *TechNewsWorld*, June 2, 2006, https://www.technewsworld.com/story/sophos-cracks-archiveus-ransomware-code-50881.html; Cary Kostka, "What Is Archiveus Trojan? A Part of the History of Modern Ransomware", Ransomware.org, February 23, 2022, https://ransomware.org/blog/archiveus-trojan-a-part-of-the-history-of-modern-ransomware/

13. Already in 1973 British mathematician Clifford Cocks created a public key algorithm, but it was kept classified by the U.K.'s GCHQ until 1997. Also see: "The Alternative History of Public-Key Cryptography", Cryptome, accessed May 4, 2024, https://cryptome.org/ukpk-alt.htm; based on Simon Singh, *The Code Book: The Science of Secrecy from Ancient Egypt to Quantum Cryptography* (Random House, 1999), 279–92.

14. TechNewsWorld, "Sophos Cracks Archiveus Ransomware Code".

15. In later years, it became common for ransomware to masquerade as law enforcement agencies, falsely accusing users of illegal activities online and demanding payment of a fine. This type of ransomware is termed 'scareware'. A prominent example is the Reveton ransomware, which emerged in 2012. It displayed a message claiming to be from United States law enforcement, alleging that the user had engaged in activities such as using pirated software or accessing child pornography. In some cases, Reveton would activate the victim's camera to suggest that a recording was being made. "Reveton Worm Ransomware", KnowBe4, accessed May 4, 2024, https://www.knowbe4.com/reveton-worm; Jonathan Reed, "How Reveton Ransomware-as-a-Service Changed Cybersecurity", *Security Intelligence*, December 19, 2022, https://securityintelligence.com/articles/how-reveton-raas-changed-cybersecurity/; Marlese Lessing, "Case Study: Reveton Ransomware", SDX Central, accessed May 4, 2024, https://www.sdxcentral.com/security/definitions/what-is-ransomware/case-study-reveton-ransomware/

16. "Archiveus Ransomware Trojan Threat Analysis", Secureworks, https://www.secureworks.com/research/arhiveus

17. Researchers from the cybersecurity firm Sophos eventually identified and disclosed the decryption key, mitigating the damage inflicted by the ransomware. Kostka, "What Is Archiveus Trojan? A Part of the History of Modern Ransomware".

18. Denis Nazarov and Olga Emelyanova, "Blackmailer: The Story of Gpcode", *SecureList*, June 26, 2006, https://securelist.com/blackmailer-the-story-of-gpcode/36089/

19. "The proud blackmailer even created a website; effectively 'RSA for dummies'", Nazarov and Emelyanova, "Blackmailer: The Story of Gpcode".

20. According to Chainalysis, since 2020, more than 90 percent of ransom payments associated with significant ransomware strains have been traced to ransomware that is deliberately programmed to exclude victims from the CIS. Chainalysis, "Eastern Europe's Crypto Crime Landscape: Scams Dominate, Plus Significant Ransomware Activity", October 14, 2021, https://www.chainalysis.com/blog/eastern-europe-cryptocurrency-geography-report-2021-preview/

21. Users found their files encrypted, and it was unclear which program had been used for encryption. The only clue left by the virus, aside from the encrypted files, was a text file named Vnimanie!.txt ('vnimanie' translates to 'attention' in Russian). This file suggested the virus's Russian origins, as both the file name and the text within were in Russian. Additionally, some of the encrypted file formats were almost exclusively used in Russia, further pointing to the source of the virus. Nazarov and Emelyanova, "Blackmailer".

22. Jonathan Lusthaus, *Industry of Anonymity: Inside the Business of Cybercrime* (Harvard University Press, 2018).

23. It was the brainchild of a Ukrainian cybercriminal known as Script, who had taken inspiration from the likes of carder.org and carder.ru. The origins of CarderPlanet are shrouded in myth, fueled in part by rumors that the key players met in person and convened a cybercrime convention in Odessa, Ukraine. Lusthaus, *Industry of Anonymity*.

24. Script was influenced by *The Godfather*. Lusthaus, *Industry of Anonymity*, p. 43–44.

25. Ryan W. Neal, "CryptoLocker Virus Holds Computers For Ransom", *International Business Times*, October 21, 2013, https://www.ibtimes.com/cryptolocker-virus-new-malware-holds-computers-ransom-demands-300-within-100-hours-threatens-encrypt; Bleeping Computer, "CryptoLocker Ransomware Information Guide and FAQ", October 14, 2013, https://www.bleepingcomputer.com/virus-removal/cryptolocker-ransomware-information

26. Kurt Baker, "History of Ransomware", CrowdStrike, October 10, 2022, https://www.crowdstrike.com/cybersecurity-101/ransomware/history-of-ransomware/

27. This was not the first time this happened. A previous instance was Jigsaw.

28. KnowBe4, "Reveton Worm Ransomware".

29. "Most Wanted: EvGeniy Mikhailovich Bogachev", FBI, accessed May 4, 2024, https://www.fbi.gov/wanted/cyber/evgeniy-mikhailovich-bogachev

30. "U.S. Leads Multi-National Action Against 'Gameover Zeus' Botnet and 'Cryptolocker' Ransomware, Charges Botnet Administrator", United States Department of Justice: Office of Public Affairs, June 2, 2014, https://www.justice.gov/opa/pr/us-leads-multi-national-action-against-gameover-zeus-botnet-and-cryptolocker-ransomware

31. For more on the market changes see: Baker, "History of Ransomware".

32. "CryptoWall Ransomware", KnowBe4, accessed May 5, 2024, https://www.knowbe4.com/cryptowall

33. Baker, "History of Ransomware".

34. RaaS was already applied for the first time in 2014, but it grew further in the following years.

35. For an overview of RaaS compared to earlier forms of ransomware, sometimes called 'commodity ransomware', also see: Kris Oosthoek, Jack Cable, Georgios Smaragdakis, "A Tale of Two Markets: Investigating the Ransomware Payments Economy", *Communications of the ACM*, 66(8), 2023, 74–83, https://doi.org/10.1145/3582489

36. Also see: Adam Marget, "Ransomware-as-a-Service (RaaS): What It Is & How It Works", *Unitrends*, August 5, 2022, https://www.unitrends.com/blog/ransomware-as-a-service-raas

37. "New Stampado Ransomware Sold Cheap on the Dark Web", *Trend Micro DE*, July 14, 2016, https://www.trendmicro.com/vinfo/de/security/news/cybercrime-and-digital-threats/new-stampado-ransomware-sold-cheap-on-the-dark-web. But it was also ineffective: Lawrence Abrams, "Stampado Ransomware Campaign Decrypted before It Started", *Bleeping Computer*, July 22, 2016, https://www.bleepingcomputer.com/news/security/stampado-ransomware-campaign-decrypted-before-it-started/

38. On the distinctive approach in ransomware distribution of Shark: Lawrence Abrams, "The Shark Ransomware Project Allows You to Create Your Own Customized Ransomware", *Bleeping Computer*, August 15, 2016, https://www.bleepingcomputer.com/news/security/the-shark-ransomware-project-allows-to-create-your-own-customized-ransomware/

39. Operators also devised engagement strategies, such as deactivating dormant affiliate accounts, to maintain a dynamic and active affiliate base. Also see: "CONTI Ransomware Group", *PRODAFT*, accessed May 5, 2024, https://resources.prodaft.com/conti-ransomware-group-report

40. Also see: Diana Granger, "Fatboy Ransomware-as-a-Service Emerges on Russian-Language Forum", *Recorded Future*, May 4, 2017, https://www.recordedfuture.com/blog/fatboy-ransomware-analysis

41. Granger, "Fatboy Ransomware-as-a-Service Emerges on Russian-Language Forum".

42. "Our Big Mac Index Shows How Burger Prices Are Changing", *The Economist*, January 25, 2024, https://www.economist.com/big-mac-index

43. Granger, "Fatboy Ransomware-as-a-Service Emerges on Russian-Language Forum".

44. Translation from Recorded Future, Granger, "Fatboy Ransomware-as-a-Service Emerges on Russian-Language Forum".

45. AdvIntel, "Digital 'Pharmacusa' II: The 'GandCrab' Phenomenon", Internet Archive: WayBackMachine, January 26, 2023, https://web.archive.org/web/20230126230909/https://www.advintel.io/post/digital-pharmacusa-ii-the-gandcrab-phenomenon

46. This argument was also presented by AdvIntel, "Digital 'Pharmacusa' II: The 'GandCrab' Phenomenon".

47. Brian Krebs, "Who's Behind the GandCrab Ransomware?", *Krebs on Security*, July 8, 2019, https://krebsonsecurity.com/2019/07/whos-behind-the-gand-crab-ransomware/

48. AdvIntel, "Digital 'Pharmacusa' II: The 'GandCrab' Phenomenon".

49. Lawrence Abrams, "Allied Universal Breached by Maze Ransomware, Stolen Data Leaked", *Bleeping Computer*, November 21, 2019, https://www.bleeping-computer.com/news/security/allied-universal-breached-by-maze-ransom-ware-stolen-data-leaked/; Sophos, "Maze Ransomware: Extorting Victims for 1 Year and Counting", *Sophos News*, May 12, 2020, https://news.sophos.com/en-us/2020/05/12/maze-ransomware-1-year-counting/

50. For the longer story, see: Dudley and Golden, *The Ransomware Hunting Team*.

51. Magno Logan et al., "The State of Ransomware: 2020's Catch-22—Security News", *Trend Micro IE*, February 3, 2021, https://www.trendmicro.com/vinfo/ie/security/news/cybercrime-and-digital-threats/the-state-of-ransom-ware-2020-s-catch-22

52. Ani Petrosyan, "Global Number of Ransomware Attacks Q1 2020-Q4 2022", Statista, August 29, 2023, https://www.statista.com/statistics/1315826/ran-somware-attacks-worldwide/; Rajeev Syal, "Ransomware Attacks in UK Have Doubled in a Year, Says GCHQ Boss", *The Guardian*, October 25, 2021, sec. UK news, https://www.theguardian.com/uk-news/2021/oct/25/ransom-ware-attacks-in-uk-have-doubled-in-a-year-says-gchq-boss. John Sakellariadis also suggests that since 2019, there has been a professionalization of ransom-ware. He points out that the clearest sign of how ransomware has transformed cybercrime markets from this period is the rise of illicit access brokers and the marketplaces in which they operate. John Sakellariadis, "Behind the Rise of Ransomware", *Atlantic Council*, August, 2022, https://www.jstor.org/stable/resrep42765

53. See McAfee Blog, parts 1–4, McAfee Labs, "McAfee ATR Analyzes Sodinokibi Aka REvil Ransomware-as-a-Service—What The Code Tells Us", *McAfee Blog*, October 2, 2019, https://www.mcafee.com/blogs/other-blogs/mcafee-labs/mcafee-atr-analyzes-sodinokibi-aka-revil-ransomware-as-a-service-what the-code-tells-us/; John Fokker, "Dismantling a Prolific Cybercriminal Empire: REvil Arrests and Reemergence", Trellix, September 29, 2022, https://www.trellix.com/blogs/research/dismantling-a-prolific-cybercriminal-empire/

54. You could also describe them as adopting a more corporate structure, with specialized roles such as dedicated HR or payroll managers being introduced.

55. Brian Krebs, "Is 'REvil' the New GandCrab Ransomware?", *Krebs on Security*, July 15, 2019, https://krebsonsecurity.com/2019/07/is-revil-the-new-gandcrab-ransomware/

56. Russian OSINT, *ЭЛИТНЫЕ ХАКЕРЫ REVIL/SODINOKIBI: $100 МИЛЛИОНОВ НА ШИФРОВАЛЬЩИКЕ?*, Youtube, October 23, 2020, https://www.youtube.com/watch?v=ZyQCQ1VZp8s; Dmitry Smilyanets, "'I Scrounged through the Trash Heaps … Now I'm a Millionaire:' An Interview with REvil's Unknown", *The Record*, March 16, 2021, https://therecord.media/i-scrounged-through-the-trash-heaps-now-im-a-millionaire-an-interview-with-revils-unknown; Lawrence Abrams, "Another Ransomware Will Now Publish Victims' Data If Not Paid", *Bleeping Computer*, December 12, 2019, https://www.bleepingcomputer.com/news/security/another-ransomware-will-now-publish-victims-data-if-not-paid/

57. Jonathan Greig, "REvil Ransomware Operators Claim Group Is Ending Activity Again, Victim Leak Blog Now Offline", *ZDNET*, October 18, 2021, https://www.zdnet.com/article/revil-ransomware-operators-claim-group-is-ending-activity-again-happy-blog-now-offline/

58. Mathew J. Schwartz, "REvil's Ransomware Success Formula: Constant Innovation", *Bank Info Security*, July 2, 2021, https://www.bankinfosecurity.com/revils-ransomware-success-formula-constant-innovation-a-16976

59. Lawrence Abrams, "Asteelflash Electronics Maker Hit by REvil Ransomware Attack", *Bleeping Computer*, April 2, 2021, https://www.bleepingcomputer.com/news/security/asteelflash-electronics-maker-hit-by-revil-ransomware-attack/

60. Lawrence Abrams, "JBS Paid $11 Million to REvil Ransomware, $22.5M First Demanded", *Bleeping Computer*, June 10, 2021, https://www.bleepingcomputer.com/news/security/jbs-paid-11-million-to-revil-ransomware-225m-first-demanded/

61. Fernando Martinez, "REvil's New Linux Version", *AT&T Cybersecurity*. Cybersecurity, July 1, 2021, https://cybersecurity.att.com/blogs/labs-research/revils-new-linux-version

62. Abrams, "Another Ransomware Will Now Publish Victims' Data If Not Paid".

63. Lawrence Abrams, "Sodinokibi Ransomware Says Travelex Will Pay, One Way or Another", *Bleeping Computer*, January 9, 2020, https://www.bleepingcomputer.com/news/security/sodinokibi-ransomware-says-travelex-will-pay-one-way-or-another/; Ionut Ilascu, "Sodinokibi Ransomware Hits Travelex, Demands $3 Million", *Bleeping Computer*, January 6, 2020, https://www.bleepingcomputer.com/news/security/sodinokibi-ransomware-hits-travelex-demands-3-million/; Anna Isaac, Caitlin Ostroff, and Bradley Hope, "Travelex Paid Hackers Multimillion-Dollar Ransom Before Hitting New Obstacles", *The Wall Street Journal*, April 9, 2020, https://web.archive.org/web/2022121

2140922/https://www.wsj.com/articles/travelex-paid-hackers-multimillion-dollar-ransom-before-hitting-new-obstacles-11586440800

64. This was a clear departure from their previous practice of freely posting stolen data. UNKN also mentioned that he contemplated engaging in personal harassment of company executives as a means to coerce payment, though it is unclear whether this approach was ever pursued. Smilyanets, "'I Scrounged through the Trash Heaps … Now I'm a Millionaire'".

65. On their shame site, REvil announced that—as GSM seemed to be unwilling to pay—each week it would auction off the data of one celebrity and after allegedly having successfully sold Trump's data, Madonna would be next. Ionut Ilascu, "REvil Ransomware Found Buyer for Trump Data, Now Targeting Madonna", *Bleeping Computer*, May 18, 2020, https://www.bleepingcomputer.com/news/security/revil-ransomware-found-buyer-for-trump-data-now-targeting-madonna/

66. "What Is REvil Ransomware?", Nomios Group, accessed May 5, 2024, https://www.nomios.com/resources/what-is-revil-ransomware/; Jessica Saavedra-Morales, "McAfee ATR Analyzes Sodinokibi Aka REvil Ransomware-as-a-Service—Crescendo", *McAfee Blog*, October 21, 2019, https://www.mcafee.com/blogs/other-blogs/mcafee-labs/mcafee-atr-analyzes-sodinokibi-aka-revil-ransomware-as-a-service-crescendo/

67. Smilyanets, "'I Scrounged through the Trash Heaps … Now I'm a Millionaire'".

68. Coveware: Ransomware Recovery First Responders, "Q3 Ransomware Demands Rise: Maze Sunsets & Ryuk Returns", November 4, 2020, https://www.coveware.com/blog/q3–2020-ransomware-marketplace-report

69. "JBS: Cyber-Attack Hits World's Largest Meat Supplier", *BBC News*, June 2, 2021, https://www.bbc.com/news/world-us-canada-57318965; Abrams, "JBS Paid $11 Million to REvil Ransomware, $22.5M First Demanded".

70. Unlike in most other cases at the time, REvil refrained from exfiltrating data before encrypting the victims' data. The Ransomware Files Podcast, *The Ransomware Files Podcast, Episode 6: Kaseya and REvil*, YouTube, April 8, 2022, https://www.youtube.com/watch?v=dO8hNhi9WmM. As it was launched just before a holiday weekend, it did not have as much impact in Europe. Thomas Kuhn, "Kaseya-Attacke: Wie der Deutsche Feierabend der Kaseya-Attacke Zum Verhängnis Wurde", *WirtschaftsWoche*, July 9, 2021, https://www.wiwo.de/technologie/digitale-welt/kaseya-attacke-wie-der-deutsche-feierabend-der-kaseya-attacke-zum-verhaengnis-wurde/27404374.html

71. In other words, I regard these ransom war groups as a category encompassing the most consequential ransomware groups.

72. Sri Lanka also suffered from a significant ransomware attack but did not disclose which group was behind the attack. On Cuba's attack against Montenegro: *Reuters*, "Montenegro blames criminal gang for cyber attacks on government", September 1, 2022, https://www.reuters.com/world/europe/montenegro-blames-criminal-gang-cyber-attacks-government-2022–08–31/; UK Parliament,

"Written evidence submitted by BAE Systems, (2022, December 1), https://committees.parliament.uk/writtenevidence/114375/pdf/; on Quantum's attack against the Dominican Republic: Jonathan Greig, "Dominican Republic refuses to pay ransom after attack on agrarian institute", *The Record*, August 26, 2022, https://therecord.media/dominican-republic-refuses-to-pay-ransom-after-attack-on-agrarian-institute; on Ransomhouse: Jonathan Greig, "Several Colombian Government Ministries Hampered by Ransomware Attack", *The Record*, September 15, 2023, https://therecord.media/colombia-government-ministries-cyberattack

73. BBC, "Critical incident over London hospitals' cyber-attack", June 4, 2024, https://www.bbc.com/news/articles/c288n8rkpvno. Later figures suggest a much higher number of medical procedures that had to be postponed, with often extremely harmful consequences to patients: Connor Jones, "Cancer patient forced to make terrible decision after Qilin attack on London hospitals", *The Register*, July 5, 2024, https://www.theregister.com/2024/07/05/qilin_impacts_patient/

74. Colonial Pipeline, Media Statement Updated: Colonial Pipeline System Disruption, Press Release, May 8, 2021, https://web.archive.org/web/20210508173736/https://www.colpipe.com/news/press-releases/media-statement-colonial-pipeline-system-disruption; Aaron Gregg, "Colonial Pipeline was shut down with worst-case scenario in mind, executives say", June 9, 2021, https://www.washingtonpost.com/business/2021/06/09/colonial-pipeline-mandiant-house-hearing/; Federal Motor Carrier Safety Administration, Regional Emergency Decleration under CFR § 390.23", May 9, 2021, https://www.fmcsa.dot.gov/sites/fmcsa.dot.gov/files/2021–05/ESC-SSC-WSC%20-%20Regional%20Emergency%20Declaration%202021–002%20-%2005–09–2021.pdf

2. THE MOB FRAMEWORK

1. On the potential (in)effectiveness of these strategies see: Allan Liska, "Is Double/Triple/Whatever Extortion Working?", Ransomware, August 1, 2021, https://ransomwaresommelier.com/p/is-doubletriplewhatever-extortion

2. Previously, an individual's decision to pay was less about the reputation of the attacking group and more about the believability of the ransomware's narrative. Messages like 'Pay to renew your license' or 'We have compromising photos of you; pay immediately' were the bait. It wasn't so much about the group's history of assisting once paid, but more a widespread approach, casting a wide net in hopes someone would fall for the trick. However, with the emergence of big game hunting, the objective extends beyond merely targeting larger entities.

3. The reputation-building efforts extend beyond the victim sphere into the cyber-criminal ecosystem. They leverage forums, dark web marketplaces, and social media to broadcast their achievements, promote their services, and underscore

the reliability of their operations. Such efforts create a buzz within the cyber-criminal community, further establishing their reputation and attracting potential collaborators. For a more general discussion on trust in the cybercriminal ecosystem, see: Jonathan Lusthaus, "Trust in the World of Cybercrime", *Global Crime* 13, no. 2 (May 2012): 71–94.

4. Florian Roth, Pasquale Stirparo, David Bizeul, Brian Bell, Ziv Chang, Joel Esler, Kristopher Bleich, et al., "APT Groups and Operations", March 2020, https://docs.google.com/spreadsheets/d/1H9_xaxQHpWaa4O_Son4Gx0YOIzlcBWMsdvePFX68EKU/edit#gid=1864660085

5. A significant challenge in collecting data on APTs is their tendency not to become 'extinct'. Intelligence experts often focus on reporting about new APTs while neglecting to share information about whether an APT has ceased its activities (which can also be challenging to confirm). This means that the public threat landscape can essentially only grow.

6. This model is also widely criticized in the APT space for being too narrow. Lockheed Martin, "Gaining the Advantage: Applying Cyber Kill Chain Methodology to Network Defense", 2015, https://www.lockheedmartin.com/content/dam/lockheed-martin/rms/documents/cyber/Gaining_the_Advantage_Cyber_Kill_Chain.pdf

7. Ransomware groups often use partial encryption to speed up their attacks. See: Cybereason Global SOC & Cybereason Security Research Teams, "Royal Rumble: Analysis of Royal Ransomware", May 5, 2024, https://www.cybereason.com/blog/royal-ransomware-analysis. For an overview see: Aleksander Milenkoski, "Encryption time flies when you're having fun the case of the exotic BlackCat ransomware", Virus Bulletin Conference, 2022, https://www.youtube.com/watch?app=desktop&v=VjxgXl6d0Go&ab_channel=VirusBulletin

8. Rob Joyce, "Disrupting State Hackers", USENIX Enigma, 2016, https://www.usenix.org/sites/default/files/conference/protectedfiles/engima2016_transcript_joyce_v2.pdf

9. Ibid.

10. For a similar definition see: "Persistence in Cybersecurity", *Huntress*, accessed May 5, 2024, https://www.huntress.com/defenders-handbooks/persistence-in-cybersecurity

11. Some ransomware groups have been found lurking in victims' networks for more than 100 days before encrypting the data.

12. This time is known to become shorter over the years—also see chapter six. John Shier et al., "The Active Adversary Playbook 2021", *Sophos News*, May 18, 2021, https://news.sophos.com/en-us/2021/05/18/the-active-adversary-playbook-2021/

13. I want to thank Charl van der Walt for mentioning this to me.

14. Max Smeets, *No Shortcuts: Why States Struggle to Develop a Military Cyber-Force* (Oxford University Press, 2022). If reading the whole book is not your thing, I address the topic also here: Max Smeets, "Building a Cyber Force Is Even

Harder Than You Thought", *War on the Rocks*, May 12, 2022, https://waron-therocks.com/2022/05/building-a-cyber-force-is-even-harder-than-you-thought/; Max Smeets, "Cyber Arms Transfer: Meaning, Limits and Implications", *Security Studies*, 31(1), 2022, 65–91.

15. The most thorough examination I have come across of how ransomware groups interact with the media is: Sophox X-Ops, "Press and pressure: Ransomware gangs and the media", *Sophos News*, December 13, 2023, https://news.sophos.com/en-us/2023/12/13/press-and-pressure-ransomware-gangs-and-the-media/

16. Ben Buchanan, *The Cybersecurity Dilemma: Hacking, Trust and Fear Between Nations* (Oxford University Press, 2017). A longer discussion of this dynamic is provided in the concluding chapter.

17. The group is particularly known for its destructive operations. For more on Lazarus: Novetta, "Operation Blockbuster: Unraveling the Long Thread of the Sony Attack," February, 2016, https://web.archive.org/web/20160226161828/https://www.operationblockbuster.com/wp-content/uploads/2016/02/Operation-Blockbuster-Report.pdf

18. It is estimated to have cost between $4 billion and $8 billion. Some have argued, however, that there have been more catastrophic and costly cyber attacks: Tom Johansmeyer, "Debunking NotPetya's Cyber Catastrophe Myth," *Binding Hook*, April 10, 2024, https://bindinghook.com/articles-binding-edge/debunking-notpetyas-cyber-catastrophe-myth/

19. Andy Greenberg, "The Ransomware Hackers Made Some Real Amateur Mistakes," *Wired*, May 15, 2017, https://www.wired.com/2017/05/wanna-cry-ransomware-hackers-made-real-amateur-mistakes/

20. Greenberg, "The Ransomware Hackers Made Some Real Amateur Mistakes."

21. Microsoft Digital Security Unit (DSU) Microsoft Threat Intelligence, "North Korean threat actor targets small and midsize businesses with H0lyGh0st ransomware," June 14, 2022, https://www.microsoft.com/en-us/security/blog/2022/07/14/north-korean-threat-actor-targets-small-and-midsize-businesses-with-h0lygh0st-ransomware/

22. Kaspersky Lab writes that Andariel is a subgroup of Lazarus. Kaspersky, "Andariel, a Lazarus subgroup, expands its attacks with new ransomware," 2022, August 9, https://www.kaspersky.com/about/press-releases/andariel-a-lazarus-subgroup-expands-its-attacks-with-new-ransomware; U.S. Department of Justice Office of Public Affairs, "North Korean Government Hacker Charged for Involvement in Ransomware Attacks Targeting U.S. Hospitals and Health Care Providers," July 25, 2024, https://www.justice.gov/opa/pr/north-korean-government-hacker-charged-involvement-ransomware-attacks-targeting-us-hospitals

23. Andy Greenberg, *Sandworm: A New Era of Cyberwar and the Hunt for the Kremlin's Most Dangerous Hackers* (Doubleday, 2019)

24. Anton Ivanov, and Orkhan Mamedov, "ExPetr/Petya/NotPetya Is a Wiper,

Not Ransomware," SecureList, June 28, 2017, https://securelist.com/expetr-petyanotpetya-is-a-wiper-not-ransomware/78902/

25. Also see: Dustin Volz, and Sarah Young, "White House Blames Russia for 'Reckless' NotPetya Cyber Attack," *Reuters*, February 16, 2018, sec. Technology, https://www.reuters.com/article/idUSKCN1FZ0PR/

26. First disclosed by Microsoft: Microsoft Threat Intelligence, "New 'Prestige' Ransomware Impacts Organizations in Ukraine and Poland," *Microsoft Security Blog*, October 14, 2022, https://www.microsoft.com/en-us/security/blog/2022/10/14/new-prestige-ransomware-impacts-organizations-in-ukraine-and-poland/

27. Gabby Roncone et al., "APT44: Unearthing Sandworm," *Mandiant*, 2024, https://services.google.com/fh/files/misc/apt44-unearthing-sandworm.pdf

28. U.S. Cybersecurity & Infrastructure Security Agency, "Iran-based Cyber Actors Enabling Ransomware Attacks on US Organizations ," 2024, August 28, https://www.cisa.gov/news-events/cybersecurity-advisories/aa24-241a

29. Microsoft Threat Intelligence, "Iran Surges Cyber-Enabled Influence Operations in Support of Hamas," Microsoft, February 26, 2024, https://www.microsoft.com/en-us/security/security-insider/intelligence-reports/iran-surges-cyber-enabled-influence-operations-in-support-of-hamas

30. For other examples of Chinese APTs using ransomware see: Microsoft Threat Intelligence, "Ransomware as a service: Understanding the cybercrime gig economy and how to protect yourself," May 9, 2022, https://www.microsoft.com/en-us/security/blog/2022/05/09/ransomware-as-a-service-understanding-the-cybercrime-gig-economy-and-how-to-protect-yourself/#DEV-0401; for financial purposes: Mandiant, "APT41, A Dual Espionage and Cyber Crime Operation," 2022, apt41-a-dual-espionage-and-cyber-crime-operation.pdf (google.com)

31. Aleksandar Milenkoski and Julian-Ferdinand Vögele, "ChamelGang & Friends: Cyberespionage Groups Attacking Critical Infrastructure with Ransomware," *SentinelLabs*, Forthcoming; On CatB file locker see: Jim Walter, "CatB Ransomware: File Locker Sharpens Its Claws to Steal Data with MSDTC Service DLL Hijacking," March 13, 2023, https://www.sentinelone.com/blog/decrypting-catb-ransomware-analyzing-their-latest-attack-methods/

32. S. Cybersecurity & Infrastructure Security Agency, "#StopRansomware: Ransomware Attacks on Critical Infrastructure Fund DPRK Malicious Cyber Activities," , February 9, 2023 , https://www.cisa.gov/news-events/cybersecurity-advisories/aa23-040a

33. Trust has various definitions. Here, it refers to the consistency between an actor's declared actions and their actual actions, underscoring predictability and reliability in behavior. Also, the discussion in this section on trust between ransomware groups and victims—necessary to overcome the ransomware trust paradox—is distinct from Lusthaus' general discussion of trust among cybercriminals. Lusthaus, "Trust in the World of Cybercrime."

34. These forms of trust are not mutually exclusive. Debra L. Shapiro, Blair H. Sheppard, Lisa Cheraskin, "Business on a Handshake", *Negotiation Journal*, 8:4 (1992): 365–377.

35. Ibid.

36. Lawrence Abrams, "Meet LostTrust ransomware—A likely rebrand of the MetaEncryptor gang", *Bleeping Computer*, October 1, 2023, https://www.bleepingcomputer.com/news/security/meet-losttrust-ransomware-a-likely-rebrand-of-the-metaencryptor-gang/

37. Shapiro, Sheppard, Cheraskin, "Business on a Handshake".

38. Thomas C. Schelling, *Strategy of Conflict*, (Harvard University Press 1960). In addition to repeated interactions, multiple engagements can also foster deterrence-based trust. When both parties interact frequently, they are less likely to betray each other in a single transaction, knowing that future beneficial transactions could be at risk.

39. Shapiro, Sheppard, Cheraskin, "Business on a Handshake".

40. Ibid.

41. For a more research on the role of insurers see the excellent work at RUSI by Jamie MacColl, Jason Nurse, and James Sullivan: https://rusi.org/explore-our-research/projects/ransomware-role-cyber-insurance

42. Negotiators might also face perverse incentives in this situation. For instance, if they earn a percentage of the negotiated ransom amount, they might signal to the ransomware group to double the ransom, offering to settle for half and splitting the difference.

43. See Darkside's ransom note: https://github.com/ThreatLabz/ransomware_notes/blob/main/darkside/darkside.txt

44. See Karma's ransom note: https://github.com/ThreatLabz/ransomware_notes/blob/main/karma/KARMA-ENCRYPTED.txt

45. Dissent, "At some point, SNAtch Team stopped being the Snatch ransomware gang. Were journalists the last to know?", *Databreaches*, September 1, 2023, https://databreaches.net/2023/09/01/at-some-point-snatch-team-stopped-being-the-snatch-ransomware-gang-were-journalists-the-last-to-know/

46. Sophox X-Ops, "Press and pressure: Ransomware gangs and the media", *Sophos News*, December 13, 2023, https://news.sophos.com/en-us/2023/12/13/press-and-pressure-ransomware-gangs-and-the-media/

47. Shapiro, Sheppard, Cheraskin, "Business on a Handshake".

48. Shapiro, Sheppard, Cheraskin, "Business on a Handshake".

49. The FAQs also include a section on "special offers for journalists and reporters", which indicates that they are open to interviews with journalists but require proof of their identity prior to an interview. In an attempt to downplay the malicious nature of their activities, 8Base also present themselves as "honest and simple pentesters" offering their victims the most favorable terms for the return of their data. On the FAQs, see: Brian Krebs, "Who's Behind the 8Base Ransomware Website?", *KrebsonSecurity*, September 18, 2023, https://kreb-

sonsecurity.com/2023/09/whos-behind-the-8base-ransomware-website/; on their background as pentesters see: Intel Cocktail, "8BASE Ransomware Group Interview: "We Are Honest and Simple Pentesters", 2024, https://intelcocktail.com/8base-interview/

50. For HelloKitty's ransom note: https://github.com/ThreatLabz/ransomware_notes/blob/main/hellokitty/%5BFile_Name%5D.README_TO_RESTORE

51. For Chilelocker's ransom note: https://github.com/ThreatLabz/ransomware_notes/blob/main/chilelocker/readme_for_unlock.txt

52. Lucian Constantin, "REvil Ransomware Explained: A Widespread Extortion Operation", *CSO Online*, November 12, 2021, https://www.csoonline.com/article/570101/revil-ransomware-explained-a-widespread-extortion-operation.html

53. The group's leader wrote in a post: "All affiliates come and go, and LockBit is eternal". Dina Temple-Raston, "Ransomware Diaries: Undercover with the Leader of LockBit", *The Record*, January 16, 2023, https://therecord.media/ransomware-diaries-undercover-with-the-leader-of-lockbit. See the Conclusion for a brief discussion of the takedown of this group led by the UK National Crime Agency.

54. For a discussion on the change in discourse, see: Chesney, and Smeets, eds. *Deter, Disrupt, or Deceive.*

55. When it does focus on non-state actors, it is about trying to understand its relationship to the state. For key literature on proxy relations see: Florian J. Egloff, *Cybersecurity and the Age of Privateering* (Oxford: Oxford University Press: 2022); Tim Maurer, *Cyber Mercenaries: The State, Hackers, and Power* (Cambridge: Cambridge University Press: 2018); Erica Borghard and Shawn Lonergan, "Can States Calculate the Risks of Using Cyber Proxies?", *Orbis* 60:3 (2016): 396; Jamie Collier, "Proxy Actors in the Cyber Domain: Implications for State Strategy", *St. Antony's International Review* 13:1 (2017).

56. Furthermore, as I elaborate in the conclusion, it also shifts how we deal with these threats—not just by improving security measures but by undermining the trust these criminal groups work so hard to establish.

3. THE MOB FRAMEWORK IN ANALYTICAL PRACTICE

1. It is better to assume bounded rationality—which is not the same as irrationality—and ransomware groups seek to 'satisfice' rather than 'optimize'. For an overview of the concept, see: Reinhard Selten, "What is Bounded Rationality?", *SFB Discussion Paper B-454*, May 1999, https://www.wiwi.uni-bonn.de/sfb303/papers/1999/b/bonnsfb454.pdf. For a pioneering study of bounded rationality in an institutional context, see: Joseph Jupille, Walter Mattli, and Duncan Snidal, *Institutional Choice and Global Commerce*, (Cambridge University Press: 2013).

2. That said, it did have a precursor brand named ABCD.

3. To access the Yanlowang leaks: https://www.th3protocol.com/2022/Yanlouwang-Leaks

4. Analyzing the leaked chats suggest that someone named 'saint' is the leader of the group. It also shows that there were various sub-groups within Yanlouwang and that each had a designated lead. For example, the lead of the development team was 'killanas', who had several people working under him. Also see Dina Temple-Raston, "The Yanluowang ransomware group in their own words", November 23, 2022, https://therecord.media/the-yanluowang-ransomware-group-in-their-own-words

5. The use of zero-day exploits in ransomware operations is uncommon, but there are examples. For example, in late January 2023, the Clop ransomware group exploited a zero-day vulnerability in the GoAnywhere Managed File Transfer software. This attack impacted around 130 organizations over a period of 10 days. See: Jonathan Greig, "Organizations slow to patch GoAnywhere MFT vulnerability even after Clop ransomware attacks", *The Record*, May 5, 2023, https://therecord.media/organizations-slow-to-patch-goanywhere-vulnerability-after-clop-attacks

6. For an overview see: Oren Biderman and Amir Sadon, "The Anatomy of a BlackCat (ALPHV) Attack", *Sygnia*, March 5, 2020, https://www.sygnia.co/blog/blackcat-ransomware/

7. In 2024, the BlackCat Ransomware Group collapsed following an apparent $22 million payment from Change Healthcare. It seems that the group's leaders orchestrated an "exit scam", deceiving their affiliates by withholding several ransomware payment commissions and abruptly terminating their operations. Brian Krebs, "BlackCat Ransomware Group Implodes After Apparent $22M Payment by Change Healthcare", March 5, 2024, https://krebsonsecurity.com/2024/03/blackcat-ransomware-group-implodes-after-apparent-22m-ransom-payment-by-change-healthcare/

8. Royal comes out of Conti and I therefore discuss the group in more detail chapter nine of the book. What the two examples show is that each approach—the in-house setup versus the outsourcing setup—offers different advantages in terms of scaling operations and adapting to new challenges, similar to conventional businesses.

9. This move to Telegram partly stemmed from the WhatsApp privacy policy controversy in January 2021: WhatsApp disclosed that it had been sharing users' sensitive data with Facebook for several years. For an overview, see: SOC Radar, "Telegram: A New Place for Hackers", March 24, 2022, https://socradar.io/telegram-a-new-place-for-hackers/

10. When it comes to selling data, organized criminal groups often turn to hacker forums on the Dark Web. They may also choose to publicly share the information—either after selling the data dump or if they fail to sell it—by posting the leaks on hacker forums, Pastebin, and Telegram. For example, on built-in chat features in site, see the 8Base ransomware group.

11. Lawrence Abrams, "Popular Russian hacking forum XSS bans all ransomware topics", May 13, 2023, *Bleeping Computer*, https://www.bleepingcomputer.

com/news/security/popular-russian-hacking-forum-xss-bans-all-ransomware-topics/; Sergiu Gatlan, "Ransomware ads now also banned on Exploit cyber-crime forum", May 14, 2021, *Bleeping Computer*, https://www.bleepingcomputer.com/news/security/ransomware-ads-now-also-banned-on-exploit-cybercrime-forum/

12. The organizational component benefits from studies of the platforms and channels a ransomware group use to communicate its messages—whether this is the dark web, public forums, social media, or dedicated websites—as does the branding component.

13. It is believed that Egregor emerged from the Maze ransomware group. The Maze gang abruptly ceased operations in September 2020, just a few weeks after Egregor began its activities: https://www.zdnet.com/article/egregor-ransomware-operators-arrested-in-ukraine/

14. Chris Caridi and Allison Wikoff, "This chat is being recorded: Egregor ransomware negotiations uncovered", *Security Intelligence*, July 21, 2021, https://securityintelligence.com/posts/egregor-ransomware-negotiations-uncovered/. Also see: Tim Starks, "Chat logs show how Egregor, an $80 million ransomware gang, handled negotiations with little mercy", *Cyberscoop*, July 21, 2021, https://cyberscoop.com/egregor-chat-logs-ibm-ransomware-negotiations/

15. Ibid.

16. Joe Tidy, "Mysterious 'Robin Hood' hackers donating stolen money", https://www.bbc.com/news/technology-54591761

17. Joseph Cox, "Pipeline Hackers Say They're 'Apolitical,' Will Choose Targets More Carefully Next Time", *Vice*, May 10, 2021, https://www.vice.com/en/article/bvzzez/colonial-pipeline-hackers-statement-darkside

18. Colin Cowie, "Yanlouwang Ransomware Leaks", Colins Security Blog, accessed May 5, 2024, https://www.th3protocol.com/2022/Yanlouwang-Leaks

19. Marcello Ienca and Effy Vayena, "Ethical requirements for responsible research with hacked data", *Nature Machine Intelligence, 3* (2021), 744–748.

20. Additionally, many researchers, particularly those in the commercial threat intelligence sector, have successfully penetrated online forums and chat servers frequented by ransomware groups, often sharing their insights with the wider community. However, the ethical implications of these actions are more complex to defend. Notably, infiltrating these networks may necessitate researchers engaging in questionable activities initially. Moreover, publicizing such infiltrations could prompt these ransomware groups to substantially alter their operational security practices and behaviors, which could complicate ongoing surveillance efforts by law enforcement agencies. See, for example: Jon DiMaggio, "Ransomware Diaries: Volume 1", Analyst1, January 16, 2023, https://analyst1.com/ransomware-diaries-volume-1/

21. Europol, "World's Most Dangerous Malware EMOTET Disrupted through Global Action", European Union Agency for Law Enforcement Cooperation, January 27, 2021, https://www.europol.europa.eu/media-press/newsroom/

news/world's-most-dangerous-malware-emotet-disrupted-through-global-action; also see Pieter Jansen, "Cyberhelden—Episodes", *Cyberhelden.nl*, April 14, 2022, https://www.cyberhelden.nl/episodes/

22. See the ransomware study of Ian W. Gray et al., "Money Over Morals: A Business Analysis of Conti Ransomware", *arXiv*, April 23, 2023, https://arxiv.org/abs/2304.11681

4. DATA

1. Robert S. Mueller, "RSA Cyber Security Conference", FBI: Federal Bureau of Investigation, March 1, 2012, https://archives.fbi.gov/archives/news/speeches/combating-threats-in-the-cyber-world-outsmarting-terrorists-hackers-and-spies

2. Due to space limitations, the book could only accommodate a 'Selected Bibliography' at the end. I have primarily included sources from chapter one and my case study in this section.

3. Sean Lyngaas, "'I Can Fight with a Keyboard': How One Ukrainian IT Specialist Exposed a Notorious Russian Ransomware Gang", CNN, March 30, 2022, https://www.cnn.com/2022/03/30/politics/ukraine-hack-russian-ransomware-gang/index.html

4. Shortly after the messages were posted, the download links on anonfiles were deactivated. However, VX Underground and other platforms stepped in, hosting the files on their own infrastructure, ensuring researchers could still access and analyze them. See: https://share.vx-underground.org/Conti/. There are parts of the data that were translated that can be downloaded under following links, which I used for this study: Jair Santanna, "The COMPLETE Translation of Leaked Files Related to Conti Ransomware Group", GitHub, accessed May 5, 2024, https://github.com/NorthwaveSecurity/complete_translation_leaked_chats_conti_ransomware; Kostas Tsialemis, "TRANSLATED Conti Leaked Comms", GitHub, 2022, https://github.com/tsale/translated_conti_leaked_comms; Ionut Ilascu, "Translated Conti Ransomware Playbook Gives Insight into Attacks", *Bleeping Computer*, September 2, 2021, https://www.bleepingcomputer.com/news/security/translated-conti-ransomware-playbook-gives-insight-into-attacks/; The Parmak, "Conti-Leaks-Englished", GitHub, accessed May 5, 2024, https://github.com/TheParmak/conti-leaks-englished

5. This is not a quote from Danylo but directly from CNN. Lyngaas, "'I Can Fight with a Keyboard': How One Ukrainian IT Specialist Exposed a Notorious Russian Ransomware Gang".

6. Lawrence Abrams, "Conti Ransomware Source Code Leaked by Ukrainian Researcher", *Bleeping Computer*, March 1, 2022, https://www.bleepingcomputer.com/news/security/conti-ransomware-source-code-leaked-by-ukrainian-researcher/

7. As Jon DiMaggio wonders, "it baffles me why Conti waited to shut down their

servers after realizing someone jeopardized their entire operation. Further, why would they continue to use the compromised infrastructure for nearly a week after its exposure?". Jon DiMaggio, "A Behind the Scenes Look into Investigating Conti Leaks", *Analyst1*, March 21, 2022, https://analyst1.com/a-behind-the-scenes-look-into-investigating-conti-leaks/

8. In chapter five I will discuss these leaked manuals in greater detail.

9. Gray et al., "Money Over Morals: A Business Analysis of Conti Ransomware". The subsequent leaks from Danylo shed light on Conti's reaction to the initial leak. These include conversations among prominent Conti managers and operators, Mango, Viper, and Ford, discussing the necessity of altering nicknames in response to the disclosures.

10. For the most in-depth exploration of the Trickbot leaks see: Joe Wrieden, "Who Is Trickbot? Analysis of the Trickbot Leaks", CYJAX: Digital Intelligence Securing the Future, July 2022, https://www.cyjax.com/wp-content/uploads/2022/07/Who-is-Trickbot.pdf

11. Some have referred to Danylo's leaks as the 'Panama Papers of cybercrime': John Fokker, and Jambul Tologonov, "Conti Leaks: Examining the Panama Papers of Ransomware", Trellix, March 31, 2022, https://www.trellix.com/en-gb/blogs/research/conti-leaks-examining-the-panama-papers-of-ransomware/

12. Misha Glenny, *Darkmarket: How Hackers Became the New Mafia* (Vintage, 2012), 105.

13. Furthermore, most of the messages are simply dull.

14. These concern components of TrickBot source code. "dero" refers to command dispatcher, and "lero" refers to a data collector. They are both built using Erlang programming language; see Lisa Vaas, "Conti Ransomware Decryptor, TrickBot Source Code Leaked—Vulnerability Database", Vulners Database, March 2, 2022, https://vulners.com/threatpost/THREATPOST:0B290DDF3FE14178760FDC2229CB1383

15. For a list, see: Chelsea @seadev@infosec.exchange on Twitter: "For those following the #ContiLeaks, here are a few additional translations: Hell = AD YES = DA wheelbarrow = host Cars = hosts Credits = creds (usrname/pw) Vmik = wmic Grid = network Facial expressions = mimikatz (1/2)", Twitter; for Python code to address some of these issues: Thomas Roccia?? ?? (@fr0gger_) / Twitter—who created: Thomas Roccia, "Using Python to Unearth a Goldmine of Threat Intelligence from Leaked Chat Logs", *Jupyter Security Break*, accessed May 5, 2024, https://jupyter.securitybreak.io/Conti_Leaks_Analysis/Conti_Leaks_Notebook_TR.html

16. On the use of personas, and the trade-offs that come with maintaining the same persona across different communication channels and fora see: Lusthaus, *Industry of Anonymity*.

17. Brian Krebs has published an excellent series of blog posts on Conti. See Brian Krebs, "Conti Ransomware Group Diaries, Part I: Evasion", *Krebs on Security*,

March 1, 2022, https://krebsonsecurity.com/2022/03/conti-ransomware-group-diaries-part-i-evasion/; Brian Krebs, "Conti Ransomware Group Diaries, Part II: The Office", *Krebs on Security*, March 2, 2022, https://krebsonsecurity.com/2022/03/conti-ransomware-group-diaries-part-ii-the-office/; Brian Krebs, "Conti Ransomware Group Diaries, Part III: Weaponry", *Krebs on Security*, March 4, 2022, https://krebsonsecurity.com/2022/03/conti-ransomware-group-diaries-part-iii-weaponry/; Brian Krebs, "Conti Ransomware Group Diaries, Part IV: Cryptocrime", *Krebs on Security*, March 8, 2022, https://krebsonsecurity.com/2022/03/conti-ransomware-group-diaries-part-iv-cryptocrime/

18. For example, see: Chuong Dong, "Conti Ransomware", *Chuong Dong*, December 15, 2020, https://cdong1012.github.io//reverse%20engineering/2020/12/15/ContiRansomware/

5. THE ORIGINS: RYUK

1. Ionut, Ilascu, "Hackers Ask for $5.3 Million Ransom, Turn Down $400k, Get Nothing", *Bleeping Computer*, September 5, 2019, https://www.bleepingcomputer.com/news/security/hackers-ask-for-53-million-ransom-turn-down-400k-get-nothing/

2. NewBedford, "Mayor Discusses Impact of Ransomware Attack on New Bedford's Computer System", May 5, 2024, https://www.newbedford-ma.gov/blog/news/mayor-discusses-impact-of-ransomware-attack-on-new-bedfords-computer-system/

3. Ryuk employs a traditional AES-RSA combination that is typically impossible to decrypt unless errors were made during its implementation. Catalin Cimpanu, "Ryuk Ransomware Crew Makes $640,000 in Recent Activity Surge", *Bleeping Computer*, August 21, 2018, https://www.bleepingcomputer.com/news/security/ryuk-ransomware-crew-makes-640-000-in-recent-activity-surge/

4. Ibid.

5. Ionut Ilascu, "US Accounts for More than Half of World's Ransomware Attacks", *Bleeping Computer*, August 8, 2019, https://www.bleepingcomputer.com/news/security/us-accounts-for-more-than-half-of-worlds-ransomware-attacks/

6. Experts from cybersecurity firm Check Point pointed to clearer associations between Hermes and Ryuk, along with clear differences. It is worth noting that this report emerged later, and Check Point did not conclude that Ryuk originated from North Korea. Check Point Research, "A Targeted Campaign Break-Down—Ryuk Ransomware", August 20, 2018, https://research.checkpoint.com/2018/ryuk-ransomware-targeted-campaign-break/

7. In 2016, a group of private industry partners, led by Novetta, released an extensive report on the Lazarus Group. This report unveiled that the group had been operational since at least 2009, potentially even dating back to 2007. Novetta, "Operation Blockbuster: Unraveling the Long Thread of the Sony Attack".

8. For more context and discussions about North Korea's cyber 'heists' see: Geoff

White, *The Lazarus Heist: From Hollywood to High Finance: Inside North Korea's Global Cyber War* (Penguin Business, 2022).

9. See reporting from BAE Systems for more detail: Sergei Shevchenko, Hirman Muhammad bin Abu Bakar, and James Wong, "BAE Systems Threat Research Blog: Taiwan Heist: Lazarus Tools and Ransomware", *BAE Systems Threat Research Blog*, October 16, 2017, https://baesystemsai.blogspot.com/2017/10/taiwan-heist-lazarus-tools.html. A write-up was also done by: Catalin Cimpanu, "North Korean Hackers Used Hermes Ransomware to Hide Recent Bank Heist", *Bleeping Computer*, October 17, 2017, https://www.bleepingcomputer.com/news/security/north-korean-hackers-used-hermes-ransomware-to-hide-recent-bank-heist/

10. Lawrence Abrams, "Ryuk Ransomware Partners with TrickBot to Gain Access to Infected Networks", *Bleeping Computer*, January 12, 2019, https://www.bleepingcomputer.com/news/security/ryuk-ransomware-partners-with-trickbot-to-gain-access-to-infected-networks/

11. Ibid.

12. Also see: Eugenio, "A Targeted Campaign Break-Down—Ryuk Ransomware".

13. Researchers identified over 40 processes and more than 180 services terminated by Ryuk.

14. Catalin Cimpanu, "Ryuk Ransomware Crew Makes $640,000 in Recent Activity Surge", *Bleeping Computer*, August 21, 2018, https://www.bleepingcomputer.com/news/security/ryuk-ransomware-crew-makes-640–000-in-recent-activity-surge/

15. Ibid.

16. For a longer write-up, also see: Anton Wendel, "Emotet Harvests Microsoft Outlook", *Cyber.WTF*, October 12, 2017, https://cyber.wtf/2017/10/12/emotet-beutet-outlook-aus/

17. Ibid.

18. Threat Hunter Team, "The Evolution of Emotet: From Banking Trojan to Threat Distributor", *Symantec Enterprise Blogs: Threat Intelligence*, July 18, 2018, http://prod-blogs-ui.client-b1.bkjdigital.com/blogs/threat-intelligence/evolution-emotet-trojan-distributor

19. TrickBot is believed to have ties to Dyreza, an earlier credential-stealing malware. They share operational and structural similarities, including communication with command-and-control servers. Kurt Baker, "What Is TrickBot Malware?" CrowdStrike, October 3, 2023, https://www.crowdstrike.com/cybersecurity-101/malware/trickbot/; on the modularity of Trickbot also see: Yelisey Boguslavskiy, "The TrickBot Saga's Finale Has Aired: Spinoff is Already in the Works", Internet Archive: WayBackMachine, March 1, 2022, https://web.archive.org/web/20220301022043/https://www.advintel.io/post/the-trickbot-saga-s-finale-has-aired-but-a-spinoff-is-already-in-the-works

20. Baker, "What Is TrickBot Malware?"; AdvIntel and Eclypsium, "Persist, Brick, Profit—TrickBot Offers New 'TrickBoot' UEFI-Focused Functionality", Internet Archive: WayBackMachine, April 16, 2022.

21. Charlotte Hammond and Ole Villadsen state that the group behind Trickbot subsequently created and managed new malware like BazarLoader, Anchor, and Bumblebee. These were employed to establish a presence in victim environments, paving the way for ransomware attacks following the development and operation of Ryuk, Conti, and Diavol ransomware operations. Charlotte Hammond, and Ole Villadsen, "The Trickbot/Conti Crypters: Where Are They Now?" *Security Intelligence*, June 27, 2023, https://securityintelligence.com/x-force/trickbot-conti-crypters-where-are-they-now/

22. For a more comprehensive overview: Intel471, "Understanding the Relationship between Emotet, Ryuk and TrickBot", April 14, 2020, https://intel471.com/blog/understanding-the-relationship-between-emotet-ryuk-and-trickbot

23. For an example of the combo, see: Lawrence Abrams, "Ryuk Ransomware Likely Behind New Orleans Cyberattack", *Bleeping Computer*, December 15, 2019, https://www.bleepingcomputer.com/news/security/ryuk-ransomware-likely-behind-new-orleans-cyberattack/

24. Intel471, "Understanding the Relationship between Emotet, Ryuk and TrickBot".

25. Ionut Ilascu, "Sodinokibi, Ryuk Ransomware Drive up Average Ransom to $111,000", *Bleeping Computer*, May 2, 2020, https://www.bleepingcomputer.com/news/security/sodinokibi-ryuk-ransomware-drive-up-average-ransom-to-111-000/

26. This is based on data from October 2020. Ionut Ilascu, "Ransomware Threat Surge, Ryuk Attacks about 20 Orgs per Week", *Bleeping Computer*, October 6, 2020, https://www.bleepingcomputer.com/news/security/ransomware-threat-surge-ryuk-attacks-about-20-orgs-per-week/

27. Ionut Ilascu, "How Ryuk Ransomware Operators Made $34 Million from One Victim", *Bleeping Computer*, November 7, 2020, https://www.bleepingcomputer.com/news/security/how-ryuk-ransomware-operators-made-34-million-from-one-victim/

28. Lawrence Abrams, "Conti Ransomware Shows Signs of Being Ryuk's Successor", *Bleeping Computer*, July 9, 2020, https://www.bleepingcomputer.com/news/security/conti-ransomware-shows-signs-of-being-ryuks-successor/

29. On the scaling down from March 2020, see: CrowdStrike Intel Team, "Wizard Spider Update: Resilient, Reactive and Resolute", *CrowdStrike Blog*, October 16, 2020, https://www.crowdstrike.com/blog/wizard-spider-adversary-update/. As a Prodraft reported notices, "The group did not establish its own website until early 2020 on the address http://fylszpcqfel7joif.onion. [..]". Data from victim organizations was shared on the Conti extortion site: https://continews.click and http://continewsnv5otx5kaoje7krkto2qbu3gtqef22mnr7eaxw3y6n-cz3ad.onion

30. "TRM Analysis Corroborates Suspected Ties Between Conti and Ryuk Ransomware Groups and Wizard Spider", *TRM Insights*, April 6, 2022, https://www.trmlabs.com/post/analysis-corroborates-suspected-ties-between-conti-and-ryuk-ransomware-groups-and-wizard-spider

31. For early discussion, see: Abrams, "Conti Ransomware Shows Signs of Being Ryuk's Successor".

32. Jackie Burns Koven, X, March 1, 2022, https://x.com/JBurnsKoven/status/1498679108812877824

33. See Gray et al., "Money Over Morals: A Business Analysis of Conti Ransomware".

34. An arrest warrant has been issued for a Trickbot collective member known as "max", whose real identity is Alla Witte. This suggests that many members of Trickbot were also involved in the Dyre Trojan, which preceded Trickbot. After Dyre's takedown in 2015, the remaining members of the Dyre collective shifted to Trickbot. See Gray et al., "Money Over Morals."; reference to arrest: "Latvian National Charged for Alleged Role in Transnational Cybercrime Organization", United States Department of Justice: Office of Public Affairs, June 4, 2021, https://www.justice.gov/opa/pr/latvian-national-charged-alleged-role-transnational-cybercrime-organization

35. After Alla Witte's arrest, Conti's leadership initiated a plan to finance her legal defense. "Mango", a mid-level manager within Conti, stated that they had enlisted a lawyer for her with significant connections and expertise in legal affairs, which could potentially grant them leverage over the investigation. While Stern endorsed the strategy, its successful execution remains uncertain. Furthermore, there is speculation that supporting Witte's legal expenses might not have been solely altruistic. Mango suggested that by doing so, Conti could potentially obtain insider information about the ongoing government probe into Trickbot. Also see write up by Brian Krebs, see: "How Does One Get Hired by a Top Cybercrime Gang?", *Krebs on Security*, June 15, 2021, https://krebsonsecurity.com/2021/06/how-does-one-get-hired-by-a-top-cybercrime-gang/

36. Also see: Lawrence Abrams, "FBI Links Diavol Ransomware to the TrickBot Cybercrime Group", *Bleeping Computer*, January 20, 2022, https://www.bleepingcomputer.com/news/security/fbi-links-diavol-ransomware-to-the-trickbot-cybercrime-group/

37. Stern also instructed to find samples from Maze.

38. "Ryuk's Return", *The DFIR Report*, October 8, 2020, https://thedfirreport.com/2020/10/08/ryuks-return/

39. Vitali Kremez tragically died during a scuba dive off the coast of Hollywood Beach in Florida; Daryna Antoniuk, "Cyber Community Mourns Renowned Researcher Vitali Kremez", *The Record*, November 3, 2022, https://therecord.media/cyber-community-mourns-renowned-researcher-vitali-kremez

40. On the Emotet side, an important person for Stern was a person known as "Mors (aka Veron)". Early on in Conti's operations, on September 21, 2020, Stern also mentioned to Mango, "mors is our most important person". Mango replied, "the most important coder or generally speaking?".

41. This is based on Chainalysis data on the top 10 ransomware strains by revenue in 2021. Also, as *The Record* notes, Conti is also notable for remaining consis-

tently active throughout 2021, with its share of all ransomware revenue increasing over the year. Adam Janofsky, "Ransomware Victims Paid More than $600 Million to Cybercriminals in 2021", *The Record*, February 10, 2022, https://therecord.media/ransomware-victims-paid-more-than-600-million-to-cyber-criminals-in-2021

6. THE PLAYBOOK

1. *Conti Cyber Attack on the HSE Independent Post: Incident Review* (PwC, 2021), 15, https://www.hse.ie/eng/services/publications/conti-cyber-attack-on-the-hse-full-report.pdf

2. Ibid, Ch. 3; Office of the Comptroller and Auditor General, "12. Financial Impact of Cyber Security Attack", 2022, https://data.gov.ie/dataset/financial-impact-of-cyber-security-attack; Sergiu Gatlan, "Conti Ransomware Also Targeted Ireland's Department of Health", *Bleeping Computer*, May 17, 2021, https://www.bleepingcomputer.com/news/security/conti-ransomware-also-targeted-irelands-department-of-health/

3. HSE NQPSD, "A Mixed Methods Analysis of the Effectiveness of the Patient Safety Risk Mitigation Strategies Following a Healthcare ICT Failure", 2022, http://hdl.handle.net/10147/631586

4. Gatlan, "Conti Ransomware Also Targeted Ireland's Department of Health".

5. They simply sent a link to where the key was stored in the negotiation chat and provided brief instructions on how to proceed with the decryption. At the same time, they maintained the threat that the stolen data would be leaked should no ransom be paid. They even declared that they would start publishing the data the following Monday, May 24. Lawrence Abrams, "Conti Ransomware gives HSE Ireland Free Decryptor, Still Selling Data", May 20, 2021, *Bleeping Computer*, https://www.bleepingcomputer.com/news/security/conti-ransomware-gives-hse-ireland-free-decryptor-still-selling-data/

6. Although Conti never published it, a significant amount of potentially sensitive patient and employee data was stolen, presenting yet another privacy-policy-related challenge to the HSE. After having figured out what data from whom was affected by the breach, the HSE started sending out letters to over 100,000 affected individuals in November 2022, informing them about what kind of data was affected in their case and what steps they could take next. "If You Received a Letter from the HSE about the Cyber-Attack", HSE.ie, accessed May 5, 2024, https://www2.hse.ie/services/cyber-attack/received-letter; Jack Horgan-Jones and Martin Wall, "HSE Cyberattack: More than 100,000 People Whose Personal Data Stolen to Be Contacted", *The Irish Times*, November 7, 2022, https://www.irishtimes.com/health/2022/11/07/over-100000-people-whose-personal-data-stolen-in-hse-cyberattack-to-be-contacted

7. Had the key not been released, this would likely have taken much longer, and it is assumed that a non-negligible share of the encrypted data would have been lost permanently. See: PwC, *Conti Cyber Attack*, 28.

8. Office of the Comptroller and Auditor General, "12. Financial Impact of Cyber Security Attack"; Gráinne Ní Aodha, "'Real Arms Race' on Defending Irish Health System against Cyber Attacks", BreakingNews.ie, February 9, 2023, https://www.breakingnews.ie/ireland/real-arms-race-on-defending-irish-health-system-against-cyber-attacks-1430272.html

9. Sabina Weston, "Irish Police Seize Conti Domains Used in HSE Ransomware Attack", ITPro, September 6, 2021, https://www.itpro.com/security/ransomware/360786/irish-police-seize-conti-domains-used-in-hse-ransomware-attack

10. The condemnation of Conti's actions by both the Irish government and the Russian Embassy also raises questions about how the group's actions affect diplomatic relations. How does Conti's reputation within the international community influence diplomatic efforts? What role do such incidents play in shaping public perceptions of cybercriminal groups? These questions, however, are addressed in chapter eight on 'Pioneering'.

11. For a more detailed discussion on the harm caused by the ransomware attack against HSE, see: Si Horne, Gareth Mott and Jamie MacColl, "Ransomware: A Life and Death Form of Cybercrime", *RUSI Commentary*, 2024, June 25, https://rusi.org/explore-our-research/publications/commentary/ransomware-life-and-death-form-cybercrime

12. For earlier write-ups on some of these stages, also see: "KELA Intelligence Report: Analysis of Leaked Conti's Internal Data", *KELA Targeted Cyber Intelligence*, March 15, 2022, https://ke-la.com/wp-content/uploads/2022/03/KELA-Intelligence-Report-ContiLeaks-1.pdf

13. For an excellent analysis of these two guides, also see: Vedere Labs, "Analysis of Conti Leaks", *Forescout*, March 2022, https://www.forescout.com/resources/analysis-of-conti-leaks/

14. For a greater deep-dive in some tooling see: "Analysis of Leaked Conti Intrusion Procedures by eSentire's Threat Response Unit (TRU)", *eSentire*, March 18, 2022, https://www.esentire.com/blog/analysis-of-leaked-conti-intrusion-procedures-by-esentires-threat-response-unit-tru; "Lessons from the Conti Leaks", *BushidoToken Threat Intel*, April 17, 2022, https://blog.bushidotoken.net/2022/04/lessons-from-conti-leaks.html

15. As a message reads in the leaks, "Found a way of buying a Zoominfo account, 2 managers for Buza, for his pricing research, the price is 2k".

16. See Vedere Labs, "Analysis of Conti Leaks".

17. This is often because updating these devices could cause problems with critical processes or lack support from vendors.

18. Insikt Group, "China's PLA Unit 61419 Purchasing Foreign Antivirus Products, Likely for Exploitation", Record Future, May 5, 2021, https://www.recordedfuture.com/blog/china-pla-unit-purchasing-antivirus-exploitation

19. The proxy was a person(a) called Ilja.

20. Affiliated groups are also known to have done call back phishing. Daksh Kapur, "Evolution of BazarCall Social Engineering Tactics", Trellix, October 6, 2022,

https://www.trellix.com/blogs/research/evolution-of-bazarcall-social-engi-neering-tactics/; Lawrence Abrams, "BazarCall Malware Uses Malicious Call Centers to Infect Victims", *Bleeping Computer*, March 31, 2021, https://www.bleepingcomputer.com/news/security/bazarcall-malware-uses-malicious-call-centers-to-infect-victims/

21. Also see: William Thomas, "WizardSpider Using Legitimate Services as Cloak of Invisibility", *CYJAX*, April 20, 2021, https://www.cyjax.com/wizardspider-using-legitimate-services-as-cloak-of-invisibility/.

22. On this subject, also see: Crystal Investigations Team, "The Conti Leaks Part One", Crystal Intelligence, March 31, 2022, https://crystalintelligence.com/investigations/the-conti-leaks-part-one/

23. Microsoft Threat Intelligence, "BazaCall: Phony Call Centers Lead to Exfiltration and Ransomware", *Microsoft Security Blog*, July 29, 2021, https://www.micro-soft.com/en-us/security/blog/2021/07/29/bazacall-phony-call-centers-lead-to-exfiltration-and-ransomware/; Kapur, "Evolution of BazarCall Social Engineering Tactics".

24. Microsoft Threat Intelligence, "BazaCall: Phony Call Centers Lead to Exfiltration and Ransomware".

25. Vlad Stolyarov and Benoit Sevens, "Exposing Initial Access Broker with Ties to Conti", Google, March 17, 2022, https://blog.google/threat-analysis-group/exposing-initial-access-broker-ties-conti/

26. More on the vulnerability: "Microsoft MSHTML Remote Code Execution Vulnerability: CVE-2021–40444 Security Vulnerability", Microsoft Security Response Center, September 7, 2021, https://msrc.microsoft.com/update-guide/vulnerability/CVE-2021-40444

27. Satnam Narang, "ContiLeaks: Chats Reveal Over 30 Vulnerabilities Used by Conti Ransomware—How Tenable Can Help", *Tenable Blog*, March 24, 2022, https://www.tenable.com/blog/contileaks-chats-reveal-over-30-vulnerabili-ties-used-by-conti-ransomware-affiliates

28. Considering its versatility, Cobalt Strike is deployed across many stages of the operational lifecycle.

29. The trickconti forum contains instructions on how to use these tools to main-tain control of a compromised system.

30. It is also delivered through installing Trickbot. Also see: Cybereason Nocturnus Team, "Dropping Anchor: From a TrickBot Infection to the Discovery of the Anchor Malware", *Cybereason*, accessed May 6, 2024, https://www.cybereason.com/blog/research/dropping-anchor-from-a-trickbot-infection-to-the-discovery-of-the-anchor-malware; Ravie Lakshmanan, "TrickBot Malware Gang Upgrades Its AnchorDNS Backdoor to AnchorMail", *The Hacker News*, March 1, 2022, https://thehackernews.com/2022/03/trickbot-malware-gang-upgrades-its.html

31. This write-up is based on: eSentire, "Analysis of Leaked Conti Intrusion Procedures by eSentire's Threat Response Unit (TRU)".

32. The operational guide instructs operators to remotely retrieve files from administrators' workstations using special tokens and either the net Windows utility or Cobalt Strike's file browser. A crucial point is emphasized: the guide warns against directly deploying a Cobalt Strike beacon onto the system to avoid raising alarms.

33. These instructions are similar to the disclosures that have surfaced from the Conti Playbook leak of 2021.

34. For more information see: "FileZilla FTP Server—FileZilla Wiki", FileZilla Wiki, https://wiki.filezilla-project.org/FileZilla_FTP_Server

35. This is based on a case study by Sophos conducted in 2021. However, research— not specific to Conti—suggests that the median time for criminals do the encryption after breaching the network has significantly decreased. Peter Mackenzie and Tilly Travers, "What to Expect When You've Been Hit with Conti Ransomware", *Sophos News*, February 16, 2021, https://news.sophos.com/en-us/2021/02/16/what-to-expect-when-youve-been-hit-with-conti-ransomware/. Secureworks, 2023 State of the Threat: A year in review, 2023, https://www.secureworks.com/-/media/Files/US/Reports/state%20of%20the%20threat/secureworks-se-2023-sott-report.ashx; Also see: Ollie Whitehouse, "Cybersecurity in 2023 and the challenges ahead", *Binding Hook*, November 7, 2023, https://bindinghook.com/uncategorized/cybersecurity-in-2023-and-the-challenges-ahead/

36. Yaara Shriebman, "To Be CONTInued? Conti Ransomware Heavy Leaks", *Cyberint*, March 9, 2022, https://cyberint.com/blog/research/contileaks/

37. Ibid.

38. Chuong Dong, "Conti Ransomware v2", https://chuongdong.com/reverse%20engineering/2020/12/15/ContiRansomware/

39. Bleeping Computer also managed to compile the source code for Conti Version 3 without any complications, but does not see it as an improvement. Lawrence Abrams, "More Conti Ransomware Source Code Leaked on Twitter out of Revenge", *Bleeping Computer*, March 20, 2022, https://www.bleepingcomputer.com/news/security/more-conti-ransomware-source-code-leaked-on-twitter-out-of-revenge/

40. Also see: Lisa Vaas, "Conti Ransomware V. 3, Including Decryptor, Leaked", Threatpost, March 21, 2022, https://threatpost.com/conti-ransomware-v-3-including-decryptor-leaked/179006/

41. Among them are Synology Active Backup for Business, StorageCraft ShadowProtect SPX, and Veeam. As I previously mentioned, Conti's operators analyze multiple sources of information, such browsing history, running processes, and authentication logs, to seek out clues that could lead them to backup solutions employed by the victim organization. For more info on this, also see: Vitali Kremez and Yelisey Boguslavskiy, "Backup 'Removal' Solutions—From Conti Ransomware with Love", Internet Archive: WayBackMachine, February 8, 2023, https://web.archive.org/web/20230208191330/https://www.advintel.io/post/backup-removal-solutions-from-conti-ransomware-with-love

42. Krebs, "Conti Ransomware Group Diaries, Part III".

43. This was first discovered by Milwaukee-based cyber intelligence firm Hold Security, posting a screenshot on X.

44. This aspect was first noted by Brian Krebs. Krebs cites a conversation on September 14, 2021: Conti's upper manager, "Revers", questioned, "They are insured for cyber risks, so what are we waiting for?" Conti employee "Grant" inquired about dealing with the insurance company, to which Revers responded, "That's not how it works. They have a coverage budget. We simply take it, and that's it".

45. "Jigsaw Ransomware", KnowBe4, accessed May 6, 2024, https://www.knowbe4.com/jigsaw-ransomware

46. However, this does *not* mean that the decryption of data always works.

47. At least this is the highest amount based on the leaks. The analysis comes from cryptocurrency analysis firm Crystal Blockchain.

48. The Department of Justice reported seized 63.7 Bitcoins valued at around $2.3 million. The FBI traced these funds to a Bitcoin address they could access using a private key. The seized Bitcoins, part of a 75-Bitcoin ransom paid to the DarkSide group, are linked to both computer intrusion and money laundering, permitting their seizure under forfeiture laws. "Department of Justice Seizes $2.3 Million in Cryptocurrency Paid to the Ransomware Extortionists Darkside", United States Department of Justice: Office of Public Affairs, June 7, 2021, https://www.justice.gov/opa/pr/department-justice-seizes-23-million-cryptocurrency-paid-ransomware-extortionists-darkside

49. "Monero Emerges as Crypto of Choice for Cybercriminals", *The Financial Times*, accessed May 6, 2024, https://www.ft.com/content/13fb66ed-b4e2–4f5f-926a-7d34dc40d8b6

50. There are several additional features that distinguish Monero from Bitcoin: Monero, "FAQ", The Monero Project, accessed May 6, 2024, https://www.getmonero.org/get-started/faq/index.html

51. However, to incentivize its use and take advantage of its privacy benefits, ransomware groups typically offer discounts for payments made in Monero, sometimes reducing the ransom amount by 5 percent to 20 percent compared to Bitcoin payments.

52. See: Lindsey O'Donnell-Welch, "The Cat and Mouse Game of Crypto Money Laundering", *Decipher*, June 17, 2022, https://duo.com/decipher/the-cat-and-mouse-game-of-ransomware-money-laundering

53. Also, one of the mixers that Conti and its predecessor Ryuk frequently used, known as Blender.io, was shut down by U.S. law enforcement.

54. Gray et al., "Money Over Morals".

55. Ibid.

56. We often overlook a significant cost for ransomware groups to launder money. For Conti, some estimated that they allocated nearly half of their earnings to activities like using cryptocurrency mixers and mules for money laundering.

See: Tristan Puech Luce and Laurenne-Sya, "Ransomware: Inside the Former CONTI Group", *RiskInsight*, July 1, 2022, https://www.riskinsight-wavestone. com/en/2022/07/ransomware-inside-the-former-conti-group/

57. On money mules: Anna Baydakova, "Ransomware Gang Extorted 725 BTC in One Attack, On-Chain Sleuths Find", *CoinDesk*, May 17, 2022, https://www. coindesk.com/layer2/2022/05/17/ransomware-gang-extorted-725-btc-in-one-attack-on-chain-sleuths-find

58. The U.S. Department of Treasury identifies Strix as Valery Sedletski, "an administrator for the Trickbot Group, including managing servers". The U.S. Department of the Treasury, "United States and United Kingdom Sanction Members of Russia-Based Trickbot Cybercrime Gang", January 9, 2023, https://home.treasury.gov/news/press-releases/jy1256

59. Carter has a lot of experience in purchasing tools and infrastructure for Stern. In another conversation to Stern, he mentions, "I'm out of bitcoins, 6 new servers, two vpn subscriptions, ipvanish subscriptions and 24 renewals. two weeks ahead of time renewals total $1130 bitcoins".

60. This raises questions about whether the PayPal account used could also be connected to a money mule or if it is verified using stolen or counterfeit ID documents.

7. ORGANIZATIONAL STRUCTURE

1. Krebs, "Conti Ransomware Group Diaries, Part II: The Office".

2. Aaron Schaffer, "Ransomware Hackers Have a New Worst Enemy: Themselves", *The Washington Post*, October 12, 2022, https://www.washingtonpost.com/politics/2022/10/12/ransomware-hackers-have-new-worst-enemy-themselves/

3. Krebs, "Conti Ransomware Group Diaries, Part IV: Cryptocrime".

4. Also see: Matt Burgess, "Unmasking Trickbot, One of the World's Top Cybercrime Gangs", *Wired*, August 30, 2023, https://www.wired.com/story/trickbot-trickleaks-bentley/; Chainalysis Team, "U.S. and U.K. Sanction 11 Members of Trickbot Ransomware Group", *Chainalysis*, September 7, 2023, https://www.chainalysis.com/blog/trickbot-ransomware-malware-sanctions-september-2023/

5. eSentire, "Analysis of Leaked Conti Intrusion Procedures by eSentire's Threat Response Unit (TRU)".

6. This is not confirmed in the available leaks, only suggested to me in interviews.

7. Microsoft Threat Intelligence, "Analyzing Attacks That Exploit the CVE-2021–40444 MSHTML Vulnerability", *Microsoft Security Blog*, September 15, 2021, https://www.microsoft.com/en-us/security/blog/2021/09/15/analyzing-attacks-that-exploit-the-mshtml-cve-2021–40444-vulnerability/

8. For excellent research on this topic, see; Check Point Research Team, "CPR Reveals Leaks of Conti Ransomware Group", *Check Point Blog*, March 11, 2022, https://blog.checkpoint.com/security/check-point-research-revels-leaks-of-conti-ransomware-group/

9. This message was written to Ford on 30 August, 2021.

10. The link between Khano and Target was made because two Conti members, Target and Bentley, discuss the recruitment post on the hacking forum XSS on Jabber. See: DiMaggio, "A Behind the Scenes Look into Investigating Conti Leaks".

11. This is for a post for reverse engineers, posted on March 31, 2020. Also see: Ibid.

12. Check Point Research, "Leaks of Conti Ransomware Group Paint Picture of a Surprisingly Normal Tech Start-Up… Sort Of", March 10, 2022, https://research.checkpoint.com/2022/leaks-of-conti-ransomware-group-paint-picture-of-a-surprisingly-normal-tech-start-up-sort-of/

13. Check Point Research, "CPR Reveals Leaks of Conti Ransomware Group".

14. Check Point Research, "Leaks of Conti Ransomware Group Paint Picture of a Surprisingly Normal Tech Start-Up… Sort Of".

15. Ibid.

16. Gray et al., "Money over Morals".

17. Collin was likely working on a blockchain-related project. Ryan seems to communicate a lot outside of Jabber, and his role within the organization is unclear.

18. This conversation took place in May 2021. Not much later, based on a conversation with Stern, in August 2021, Salamandra seems to be close to a burnout and says he/she requires a vacation. More generally, many employees seemed to be afraid of punishment or even expulsion, as can be seen from some reactions: "I'm not in Russia and we have a flood. I suspect there will be no internet or electricity tomorrow. So I'm warning you in advance. I hope all will be ok, but I may be out of touch for 72 hours. Please do not punish me for this".

8. BUSINESS EXPANSION

1. Dor Neemani and Asaf Rubinfeld, "Diavol—A New Ransomware Used By Wizard Spider?", *Fortinet*, July 1, 2021, https://www.fortinet.com/blog/threat-research/diavol-new-ransomware-used-by-wizard-spider

2. Abrams, "FBI Links Diavol Ransomware to the TrickBot Cybercrime Group".

3. Neemani and Rubinfeld, "Diavol—A New Ransomware Used By Wizard Spider?"

4. Ionut Ilascu, "Diavol Ransomware Sample Shows Stronger Connection to TrickBot Gang", *Bleeping Computer*, August 18, 2021, https://www.bleepingcomputer.com/news/security/diavol-ransomware-sample-shows-stronger-connection-to-trickbot-gang/; Charlotte Hammond and Chris Caridi, "Analysis of Diavol Ransomware Reveals Possible Link to TrickBot Gang", *Security Intelligence*, August 17, 2021, https://securityintelligence.com/posts/analysis-of-diavol-ransomware-link-trickbot-gang/; also see binary defense, "New Ransomware 'Diavol' Being Dropped by Trickbot", *Binary Defense*, April 18, 2023, accessed May 6, 2024, https://www.binarydefense.com/resources/threat-watch/new-ransomware-diavol-being-dropped-by-trickbot/

5. Abrams, "FBI Links Diavol Ransomware to the TrickBot Cybercrime Group".

6. Ibid.

7. Harry Igor Ansoff, *Corporate Strategy* (McGraw-Hill, New York, NY, 1965).

8. On 2020 figures: Coveware: Ransomware Recovery First Responders, "Q3 Ransomware Demands Rise: Maze Sunsets & Ryuk Returns". At the time, Sodinokibi/REvil was leading, followed by Maze.

9. INSIKT Group, "Latin American Governments Targeted By Ransomware", *Recorded Future*, June 14, 2022, https://www.recordedfuture.com/blog/latin-american-governments-targeted-by-ransomware

10. Vaas, "Conti Ransomware V. 3, Including Decryptor, Leaked".

11. While I did not come across any concrete evidence, it is also plausible that Diavol's launch might have also been a calculated competitive maneuver aimed at seizing market share from competitors such as Lockbit, REvil, and Pysa. The criminal landscape is marked by intense rivalry, and the introduction of a fresh brand could position Conti to entice new affiliates and associates who had previously aligned themselves with these rival groups.

12. The links between Stern and BlackByte are less evident. In fact, it is likely that the development of BlackByte was not driven by a desire from Stern to be more resilient, but by other members of the group. It is likely a breakaway from the core group.

13. Rodel Mendrez and Lloyd Macrohon, "BlackByte Ransomware—Pt. 1 In-Depth Analysis", *Spiderlabs Blog*, Trustwave, October 15, 2021, https://www.trustwave.com/en-us/resources/blogs/spiderlabs-blog/blackbyte-ransomware-pt-1-in-depth-analysis/

14. Trend Micro Research, "Ransomware Spotlight: BlackByte", Trend, July 5, 2022, https://www.trendmicro.com/vinfo/us/security/news/ransomware-spotlight/ransomware-spotlight-blackbyte

15. Office of the Maine Attorney General, "Data Breach Notifications", August 9, 2022, https://apps.web.maine.gov/online/aeviewer/ME/40/bd184cdd-5347-4eae-92a1-63de2dcc6c2f.shtml?bd184cdd-5347-4eae-92a1-63de2dcc6c2f=breach

16. In August 2022, BlackByte was back with a new version of their operation including a new data leak site utilizing new extortion techniques borrowed from LockBit. Abrams, "BlackByte Ransomware Gang Is Back with New Extortion Tactics".

17. CISA: Cybersecurity and Infrastructure Security Agency, "Karakurt Data Extortion Group", December 12, 2023, https://www.cisa.gov/news-events/cybersecurity-advisories/aa22-152a; "Karakurt Extortion Group: Threat Profile", *Threat Intelligence*, ThreatDown by Malwarebytes, June 14, 2022, https://www.threatdown.com/blog/karakurt-extortion-group-threat-profile/

18. CISA: Cybersecurity and Infrastructure Security Agency, "Karakurt Data Extortion Group".

19. Arctic Wolf, "The Karakurt Web: Threat Intel and Blockchain Analysis", *Arctic Wolf*, April 15, 2022, https://arcticwolf.com/resources/blog/karakurt-web/

20. BleepingComputer, X, April 15, 2022, https://x.com/BleepinComputer/status/1515021245531795464

21. Ionut Ilascu, "Karakurt Revealed as Data Extortion Arm of Conti Cybercrime Syndicate", *Bleeping Computer*, April 15, 2022, https://www.bleepingcomputer.com/news/security/karakurt-revealed-as-data-extortion-arm-of-conti-cyber-crime-syndicate/

22. Ibid.

23. Ibid.

24. For excellent analyses that cover parts of this subject, see: CPR, "Leaks of Conti Ransomware Group Paint Picture of a Surprisingly Normal Tech Start-Up... Sort Of."; BushidoToken Threat Intel, "Lessons from the Conti Leaks."; Krebs, "Conti Ransomware Group Diaries, Part IV."; Fokker and Tologonov, "Conti Leaks."; Marco Figueroa, Napoleon Bing, and Bernard Silvestrini, "The Conti Leaks Insight into a Ransomware Unicorn", Internet Archive: WayBackMachine, December 7, 2023, https://web.archive.org/web/20231207134956/https://www.breachquest.com/blog/conti-leaks-insight-into-a-ransomware-unicorn/

25. Also see: Betsy Bevilacqua, "The Law Is Finally Catching Up With Ransomware Criminals", *Wired*, February 21, 2022, https://www.wired.com/story/law-fighting-ransomware-criminals/; Matt Burgess, "The Big, Baffling Crypto Dreams of a $180 Million Ransomware Gang", *Wired*, March 17, 2022, https://www.wired.com/story/conti-ransomware-crypto-payments/

26. Daniel Van Boom, "Forget Bitcoin: Inside the Insane World of Altcoin Cryptocurrency Trading", *CNET*, April 13, 2021, https://www.cnet.com/personal-finance/crypto/features/beyond-bitcoin-the-wild-world-of-altcoin-cryptocurrency-trading/; TRM Insights, "TRM Analysis Corroborates Suspected Ties Between Conti and Ryuk Ransomware Groups and Wizard Spider".

27. Figueroa, Bing, and Silvestrini, "The Conti Leaks Insight into a Ransomware Unicorn".

28. As Krebs notes, this type of competition has been proven successful in other cases. It often serves as a simple and cheap tool to acquire intellectual property for ongoing projects. Krebs, "Conti Ransomware Group Diaries, Part IV".

29. Figueroa, Bing, and Silvestrini, "The Conti Leaks Insight into a Ransomware Unicorn".

30. TRM Insights, "TRM Analysis Corroborates Suspected Ties Between Conti and Ryuk Ransomware Groups and Wizard Spider".

31. Lawrence Abrams, "Popular Russian Hacking Forum XSS Bans All Ransomware Topics", *Bleeping Computer*, May 13, 2021, https://www.bleepingcomputer.com/news/security/popular-russian-hacking-forum-xss-bans-all-ransomware-topics/

32. Quotes from: Jessica Davis, "Hackers Hate Ransomware, but It's Quickly Becoming the New DDoS", *Healthcare IT News*, September 20, 2017, https://www.healthcareitnews.com/news/hackers-hate-ransomware-its-quickly-becoming-new-ddos; "How Ransomware Has Become an 'Ethical' Dilemma in

the Eastern European Underground", *Flashpoint*, September 20, 2017, https://flashpoint.io/blog/ransomware-ethical-dilemma-eastern-european-underground. Also see: Tom Field, "How Ransomware Groups Respond to External Pressure", *Data Breach Today*, August 28, 2023, https://www.databreachtoday.com/how-ransomware-groups-respond-to-external-pressure-a-22931

33. Also see: Fokker and Tologonov, "Conti Leaks: Examining the Panama Papers of Ransomware".

34. WeChat's extensive capabilities, ranging from social interactions to financial transactions, have made it integral to the online experiences of many Chinese users. While Conti's social forum concept might not mirror WeChat's magnitude, it is an intriguing exploration of a multifaceted platform.

35. BushidoToken Threat Intel, "Lessons from the Conti Leaks".

36. "Troubled Dark Web Carding Market Loses Another Key Vendor as FBI Seizes SSNDOB", *Elliptic*, June 8, 2022, https://www.elliptic.co/blog/troubled-dark-web-carding-market-loses-another-key-vendor-as-fbi-seizes-ssndob

9. PIONEERING

1. O'Sullivan, Edkins, and Powell, "Cyber Raid Hits High Society Jeweller Graff".

2. Conti, "Graff Diamonds", October 10, 2021, Internet Archive: WayBack Machine, https://web.archive.org/web/20211010114944/https:/continews.click/

3. Kevin O'Sullivan, "Massive Cyber Heist Rocks High Society Jeweller Graff", *Mail Online*, October 30, 2021, https://www.dailymail.co.uk/news/article-10148265/Massive-cyber-heist-rocks-high-society-jeweller-Graff.html

4. O'Sullivan, Edkins, and Powell, "Cyber Raid Hits High Society Jeweller Graff".

5. Richard Chiu, "Crime Gang Apologises to Graff Jewellers over Data Leak", *Jeweller*, November 8, 2021, https://www.jewellermagazine.com/Article/10138/Crime-gang-apologises-to-Graff-Jewellers-over-data-leak

6. Ibid.

7. Ibid.

8. See e.g. Kevin O'Sullivan, "Cyber Hackers Who Carried out Jewellers Heist Make Grovelling Apology", *Mail Online*, November 6, 2021, https://www.dailymail.co.uk/news/article-10172879/Russian-cyber-hackers-carried-virtual-heist-jewellers-Graff-make-grovelling-apology.html

9. The reason for this revelation being that Graff was now suing its insurance company for not covering the paid ransom. Olivia Fletcher, "Billionaire's Jeweler Pays $7.5 Million Crypto Ransom to Hackers", *Bloomberg*, July 6, 2022, https://www.bloomberg.com/tosv2.html?vid=&uuid=204721ac-0bcf-11ef-95cf-cb8c94120d1c&url=L25ld3MvYXJ0aWNsZXMvMjAyMi0wNy0wNi9ia WxsaW9uYWlyZS1zLWpld2VsZXItcGF5cy03LTUtbWlsbGlvbi1jcnlwdG8 tcmFuc29tLXRvLWhhY2tlcnM/bGVhZFNvdXJjZT11dmVya WZ5U2JTaXwd2FsbA==

10. Fletcher, "Billionaire's Jeweler Pays $7.5 Million Crypto Ransom to Hackers".

11. Jason Healey, "The Spectrum of National Responsibility for Cyberattacks", *The Brown Journal of World Affairs* 18, no. 1 (2011): 57–70.

12. This is based on a conversation from September 22, 2020.

13. The attack retriggered discussions on ransomware between U.S. President Joseph Biden and Russia's President Vladimir Putin. The Geneva Summit in June 2021; Ellen Nakashima, and Dalton Bennett, "A Ransomware Gang Shut down after Cybercom Hijacked Its Site and It Discovered It Had Been Hacked", *The Washington Post*, November 3, 2021, https://www.washingtonpost.com/national-security/cyber-command-revil-ransomware/2021/11/03/528e03e6–3517–11ec-9bc4–86107e7b0ab1_story.html. Jon DiMaggio, "A History of REvil", *Anaylst1*, accessed May 6, 2024, https://analyst1.com/history-of-revil/. On how it triggered discussions: Steve Holland and Andrea Shalal, "Biden Presses Putin to Act on Ransomware Attacks, Hints at Retaliation", *Reuters*, July 10, 2021, sec. Technology, https://www.reuters.com/technology/biden-pressed-putin-call-act-ransomware-attacks-white-house-2021–07–09/; Georges De Moura, and Tal Goldstein, "What the Biden-Putin Summit Reveals about Future of Cyber Attacks—and How to Increase Cybersecurity", *World Economic Forum*, June 17, 2021, https://www.weforum.org/agenda/2021/06/joe-biden-vladimir-putin-summit-cybersecurity/

14. Lawrence Abrams, "REvil Ransomware Gang's Web Sites Mysteriously Shut Down", *Bleeping Computer*, July 13, 2021, https://www.bleepingcomputer.com/news/security/revil-ransomware-gangs-web-sites-mysteriously-shut-down; Nakashima and Bennett, "A Ransomware Gang Shut down after Cybercom Hijacked Its Site and It Discovered It Had Been Hacked"; Lawrence Abrams, "REvil Ransomware Is Back in Full Attack Mode and Leaking Data", *Bleeping Computer*, September 11, 2021, https://www.bleepingcomputer.com/news/security/revil-ransomware-is-back-in-full-attack-mode-and-leaking-data/

15. DiMaggio, "A History of REvil", p. 36.

16. Ellen Nakashima and Rachel Lerman, "FBI Held Back Ransomware Decryption Key from Businesses to Run Operation Targeting Hackers", *The Washington Post*, September 21, 2021, https://www.washingtonpost.com/national-security/ransomware-fbi-revil-decryption-key/2021/09/21/4a9417d0-f15f-11eb-a452-4da5fe48582d_story.html; Jonathan Greig, "FBI Decision to Withhold Kaseya Ransomware Decryption Keys Stirs Debate", *ZDNET*, September 24, 2021, https://www.zdnet.com/article/fbi-decision-to-withhold-kaseya-ransomware-decryption-keys-stirs-debate/

17. Lawrence Abrams, "REvil Ransomware's Servers Mysteriously Come Back Online", *Bleeping Computer*, September 7, 2021, https://www.bleepingcomputer.com/news/security/revil-ransomwares-servers-mysteriously-come-back-online/; Abrams, "REvil Ransomware Is Back in Full Attack Mode and Leaking Data".

18. Lawrence Abrams, "REvil Ransomware Shuts down Again after Tor Sites Were Hijacked", *Bleeping Computer*, October 17, 2021, https://www.bleepingcom-

puter.com/news/security/revil-ransomware-shuts-down-again-after-tor-sites-were-hijacked/

19. Nakashima and Bennett, "A Ransomware Gang Shut down after Cybercom Hijacked Its Site and It Discovered It Had Been Hacked".

20. Ellen Nakashima and Dalton Bennett, "Ring of Ransomware Hackers Targeted by Authorities in United States and Europe", *The Washington Post*, November 11, 2021, https://www.washingtonpost.com/national-security/revil-ransomware-arrests-doj/2021/11/08/9432dfc2-409f-11ec-a88e-2aa4632af69b_story.html

21. Yaroslav Vasinskyi, a key player in the Kaseya attack, was among those arrested. Ionut Ilascu, "Russia Arrests REvil Ransomware Gang Members, Seize $6.6 Million", *Bleeping Computer*, January 14, 2022, https://www.bleepingcomputer.com/news/security/russia-arrests-revil-ransomware-gang-members-seize-66-million/; Sergiu Gatlan, "REvil Ransomware Affiliates Arrested in Romania and Kuwait", *Bleeping Computer*, November 8, 2021, https://www.bleeping-computer.com/news/security/revil-ransomware-affiliates-arrested-in-romania-and-kuwait/

22. As *The New York Times* notes, these arrests came on the same day when the U.S. government accused Russia of dispatching saboteurs to Ukraine in order to establish grounds for invasion, and when hackers caused the shutdown of numerous Ukrainian government websites. Ivan Nechepurenko, "Russia Says It Shut Down Notorious Hacker Group at U.S. Request", *The New York Times*, January 14, 2022, sec. World, accessed May 7, 2024, https://www.nytimes.com/2022/01/14/world/europe/revil-ransomware-russia-arrests.html

23. FSB of Russia, "Illegal Activities of Members of the Organized Criminal Community Were Suppressed", Federal Security Service of the Russian Federation, January 14, 2022, http://www.fsb.ru/fsb/press/message/single.htm%21id%3D10439388%40fsbMessage.html

24. Ibid.

25. Ibid.

26. "REvil Ransomware Gang Arrested in Russia", *BBC News*, January 14, 2022, https://www.bbc.com/news/technology-59998925

27. From Khodjibaev's Twitter. Arielle Waldman, "Distrust, Feuds Building among Ransomware Groups", TechTarget: Security, February 3, 2022, https://www.techtarget.com/searchsecurity/news/252512902/Distrust-feuds-building-among-ransomware-groups

28. Waldman, "Distrust, Feuds Building among Ransomware Groups".

29. This already occurred in July, after REvil went offline for the first time. As one person, Fish, writes in the group in July of 2021: "scary things are happening", referencing public reporting on the arrest of REvil ransomware hackers. Also see: David E. Sanger, "Russia's Most Aggressive Ransomware Group Disappeared. It's Unclear Who Made That Happen", *The New York Times*, July 13, 2021, sec. U.S., https://www.nytimes.com/2021/07/13/us/politics/russia-hacking-ransomware-revil.html

30. These reasons, however, are not discussed in detail in the Jabber chats that were leaked. It is expected that they alluded to REvil overreaching and going for Kaseya.

31. It is unclear if they actually left the business.

32. Some Conti members expressed worries about their professional activities and potential consequences, even within Russia. An example is a message from Kagas, an HR department member, to Stern in early November 2021: "Hi, again, it's before the 13th. We thought we were being followed this morning, as there were unfamiliar cars in the yard, two persons sitting in the car. So as not to set anyone up, not you and not us. I and Dorirus decided to go into hiding for a week, all the devices are temporarily in the headquarters apartment. We'll be gone for about a week. It's better to be reassured because our case is extended until November 13. By November 13, everything will be solved, we hope that the case will be dropped. Since it is essentially already impossible to get anything from us. The only thing that bothers us is that the Canadians have a request for four Russians. I hope we are not on the list".

33. John Fokker and Jambul Tologonov, "Conti Leaks: Examining the Panama Papers of Ransomware", *Trellix*, March 31, 2022, https://www.trellix.com/en-gb/blogs/research/conti-leaks-examining-the-panama-papers-of-ransomware/

34. William James and Steve Scherer, "Russia Trying to Steal COVID-19 Vaccine Data, Say UK, U.S. and Canada", *Reuters*, July 16, 2020, https://www.reuters.com/article/idUSKCN24H232/; also see: Amy Walker, "UK '95% Sure' Russian Hackers Tried to Steal Coronavirus Vaccine Research", *The Guardian*, July 17, 2020, https://www.theguardian.com/world/2020/jul/17/russian-hackers-steal-coronavirus-vaccine-uk-minister-cyber-attack; Matt Burgess, "Leaked Ransomware Docs Show Conti Helping Putin From the Shadows", *Wired*, March 18, 2022, https://www.wired.com/story/conti-ransomware-russia/

35. Christo Grozev, X, February 28, 2022, https://x.com/christogrozev/status/1498386621657493510; for background: Damian Whitworth, "How I Exposed Alexei Navalny's Poisoners and Fell Foul of Putin", *The New York Times*, March 4, 2023, https://www.thetimes.co.uk/article/bellingcat-eliot-higgins-alexei-navalny-poisoning-putin-russia-ukraine-jvsmpqmdb

36. The translated text is a bit unspecific here and is susceptible to diverse interpretation. The translated text is: "understand. if they decipher it there—I will beacon".

37. Bellingcat has frequently been targeted by Russian hacking groups. In the summer of 2019, it became the victim of an extended phishing campaign that mimicked the email service ProtonMail. Bellingcat believed that this effort was orchestrated by Fancy Bear, a hacking group with ties to the Russian state. Christo Grozev, along with journalists Roman Dobrokhotov and Mark Burnett—who had collaborated on an investigation into Russian military intel-

ligence activities—were also victims of this phishing attack. Bellingcat Investigation Team, "Guccifer Rising? Months-Long Phishing Campaign on ProtonMail Targets Dozens of Russia-Focused Journalists and NGOs", *Bellingcat*, August 10, 2019, https://www.bellingcat.com/news/uk-and-europe/2019/08/10/guccifer-rising-months-long-phishing-campaign-on-protonmail-targets-dozens-of-russia-focused-journalists-and-ngos; Zak Doffman, "Russia Linked To Cyberattacks On Bellingcat Researchers Investigating GRU (Updated)", *Forbes*, July 26, 2019, https://www.forbes.com/sites/zakdoffman/2019/07/26/russian-intelligence-cyberattacked-journalists-hacking-encrypted-email-accounts/

38. It is interesting to see how different publications describe the connections between Conti and Russian entities, alluding to some links but highlighting their limited nature. The interpretations vary subtly but significantly in terms of whether there is ongoing cooperation, the informal nature of these ties, and their extent. *The Intercept* states, "Conti appears to be an independent criminal enterprise without formal ties to the Russian government". *Cyberint* comments, "When it comes to the relationship with the government, Conti is in a position in which the Russian government is not cooperating with them, but they are aware that they might be one day". *Wired* notes, "links to the FSB and Cozy Bear hackers appear ad hoc". While there may be no formal documentation of the ties between Conti and the Russian government, it is clear that there is a broadly recognized understanding among Conti's leadership about potential coopera-tion when necessary. Indeed, contrary to what *Cyberint* suggests, leaks indicate that Conti does collaborate with the FSB, making *Wired*'s assessment the most accurate. Conti is only known to have relationships with the FSB and not with the GRU (e.g., Sandworm) for more destructive targeting. See: Micah Lee, "Leaked Chats Show Russian Ransomware Gang Discussing Putin's Invasion of Ukraine", *The Intercept*, March 14, 2022, https://theintercept.com/2022/03/14/russia-ukraine-conti-russian-hackers/; Shriebman, "To Be CONTInued? Conti Ransomware Heavy Leaks"; Burgess, "Leaked Ransomware Docs Show Conti Helping Putin From the Shadows".

10. DECLINE AND REVIVAL

1. Considering that this faction also copies the techniques, tactics and procedures from Conti, they arguably do not deserve their own name and category.

2. RAMP then stood for Russian Anonymous MarketPlace. Its name came from an older forum that was discontinued until around 2017. RAMP was introduced in July 2021, attracting considerable interest from both researchers and cybercrim-inals. The forum surfaced on a domain that had previously been used by the Babuk ransomware data leak site and subsequently the Payload.bin leak site. KELA ana-lyzed the content of this new site and evaluated its potential for success. Arianne Bleiweiss, "New Russian-Speaking Forum—A New Place for RaaS?", *KELA Cyber*

Threat Intelligence, July 28, 2021, https://www.kelacyber.com/new-russian-speaking-forum-a-new-place-for-raas/; vx-underground, X, July 12, 2021, https://x.com/vxunderground/status/1414588622670532616

3. Counter Threat Unit Research Team, "GOLD ULRICK Continues Conti Operations despite Public Disclosures", Secureworks, April 21, 2022, https://www.secureworks.com/blog/gold-ulrick-continues-conti-operations-despite-public-disclosures

4. Shriebman, "To Be CONTInued? Conti Ransomware Heavy Leaks".

5. Coveware, "Conti Ransomware Recovery, Payment & Decryption Statistics", March 26, 2021, https://web.archive.org/web/20210326173558/https://www.coveware.com/conti-ransomware,

   May 14, 2021, https://web.archive.org/web/20210514090748/https://www.coveware.com/conti-ransomware,

   August 17, 2021, https://web.archive.org/web/20210817013440/https://www.coveware.com/conti-ransomware,

   February 15, 2022, https://web.archive.org/web/20220215054011/https://www.coveware.com/conti-ransomware,

   March 21, 2022, https://web.archive.org/web/20220321162225/https://www.coveware.com/conti-ransomware,

   April 22, 2022, https://web.archive.org/web/20220422132230/https://www.coveware.com/conti-ransomware

   May 7, 2024, https://www.coveware.com/conti-ransomware

6. "Conti Ransomware Group Internal Chats Leaked", *Rapid7 Blog*, March 1, 2022, https://www.rapid7.com/blog/post/2022/03/01/conti-ransomware-group-internal-chats-leaked-over-russia-ukraine-conflict/

7. Bogusalvskiy and Kremez, "DisCONTInued: The End of Conti's Brand Marks New Chapter For Cybercrime Landscape"; also see: Lawrence Abrams, "Conti Ransomware Shuts down Operation, Rebrands into Smaller Units", *Bleeping Computer*, May 19, 2022, https://www.bleepingcomputer.com/news/security/conti-ransomware-shuts-down-operation-rebrands-into-smaller-units/

8. Bogusalvskiy and Kremez, "DisCONTInued: The End of Conti's Brand Marks New Chapter For Cybercrime Landscape".

9. Ibid.

10. Shriebman, "To Be CONTInued? Conti Ransomware Heavy Leaks".

11. National Crime Agency, "Ransomware, Extortion and the Cyber Crime Ecosystem", September 11, 2023, https://nationalcrimeagency.gov.uk/who-we-are/publications/672-ransomware-extortion-and-the-cyber-crime-ecosystem/file

12. They also benefited from other information now in the public domain, such as the leaking of the BazarBackdoor APIs and the TrickBot command and control server source code; "Latest HiddenTear News", *Bleeping Computer*, accessed May 7, 2024, https://www.bleepingcomputer.com/tag/hiddentear/; Lawrence Abrams, "Babuk Ransomware's Full Source Code Leaked on Hacker Forum",

Bleeping Computer, September 3, 2021, https://www.bleepingcomputer.com/news/security/babuk-ransomwares-full-source-code-leaked-on-hacker-forum/; Lawrence Abrams, "Leaked Babuk Locker Ransomware Builder Used in New Attacks", *Bleeping Computer*, June 30, 2021, https://www.bleeping-computer.com/news/security/leaked-babuk-locker-ransomware-builder-used-in-new-attacks/

13. ScareCrow, Meow, and Putin ransomware all use the same tactic of encrypting users' files and appending unique extensions (.CROW, .MEOW, and .PUTIN, respectively) to the affected files. Each variant's ransom note instructs victims to contact the threat actors via Telegram. In addition, Putin ransomware threat actors use their Telegram channel to publish victim details. All three ransomware types use Telegram as their primary communication channel. Despite their low level of sophistication, these attacks often remain effective. "New Ransomware Strains Emerging From Leaked Conti's Source Code", *Cyble*, December 22, 2022; Shunichi Imano and Fred Gutierrez, "Ransomware Roundup—New Vohuk, ScareCrow, and AERST Variants", *FortiGuard Labs Threat Research*, Fortinet, December 8, 2022, https://www.fortinet.com/blog/threat-research/ransomware-roundup-new-vohuk-scarecrow-and-aerst-variants; "Several New Ransomware Variants Use the Leaked Conti Source Code", *BroadCOM*, December 22, 2022, https://www.broadcom.com/support/security-center/protection-bulletin/several-new-ransomware-variants-use-the-leaked-conti-source-code

14. "Protection Bulletins", BroadCOM, accessed May 7, 2024, https://www.broadcom.com/support/security-center/protection-bulletin#blt 27d34ae13f86e8aa_en-us;%20Cyble%20-%20New%20Ransomware%20 Strains%20Emerging%20From%20Leaked%20Conti%E2%80%99s%20 Source%20Code; "New Ransomware Strains Emerging From Leaked Conti's Source Code", *Cyble*, December 22, 2022, https://cyble.com/blog/new-ransomware-strains-emerging-from-leaked-contis-source-code/

15. This version is based on the BlackMatter ransomware group's source code.

16. Anuj Soni and Ryan Chapman, "The Curious Case of 'Monti' Ransomware: A Real-World Doppelganger", *BlackBerry Blog*, September 7, 2022, https://blogs.blackberry.com/en/2022/09/the-curious-case-of-monti-ransomware-a-real-world-doppelganger

17. MalwareHunterTeam, X, June 30, 2022, https://x.com/malwrhunterteam/status/1542595315915710465

18. One might question whether Monti was an 'offshoot' rather than a 'copycat'. The reason why it is very likely a copycat is because Monti lacked access to version 3 of the Conti locker, as revealed by the differing nonce values used in the ChaCha8 algorithm.

19. There was one more difference compared to Conti. The group used a commercial, cloud-based RMM platform known as Action1, which had not been previously employed in a ransomware attack. Typically, ransomware actors, includ-

ing Conti, use commercial RMMs like AnyDesk in their operations. Notably, the installation and configuration instructions for the AnyDesk RMM are explicitly outlined in the "CobaltStrike MANUALS_V2 Active Directory" attack manual, which was leaked from the Conti group in 2021.

20. Anonymous, X, February 26, 2022, https://x.com/YourAnonRiots/status/1497485333985857538; NB65, X, February 26, 2022, https://x.com/xxNB65/status/1497446425004810246. In one tweet, they claim to have "Friends and family sharing your struggles" referring to the struggles of the people in Ukraine. NB65, X, March 29, 2022, https://x.com/xxNB65/status/1508634761501237248; NB65, X, April 11, 2022, https://x.com/xxNB65/status/1513593777759428624

21. The group stated that they gained access to the company control systems, "leaking gas", "turn[ing] off fans", delet[ing] configurations/profiles", "access[ing] the file system", and ultimately causing a shutdown. CyberKnow, X, February 27, 2022, https://x.com/Cyberknow20/status/1498024125151199232

22. NB65, X, February 27, 2022, https://x.com/xxNB65/status/1497722423503949824; https://x.com/xxNB65/status/1498013842948993025

23. NB65, X, February 28, 2022, https://x.com/xxNB65/status/1498216238526255104; NB65, X, March 1, 2022, https://x.com/xxNB65/status/1498563301525102594; NB65, X, March 3, 2022, https://x.com/xxNB65/status/1499468274332942355; https://x.com/xxNB65/status/1504933046734233600; Joseph Cox, "Hackers Breach Russian Space Research Institute Website", *Vice*, March 3, 2022, https://www.vice.com/en/article/z3n8ea/hackers-breach-russian-space-research-institute-website; NB65, X, March 25, 2022, https://x.com/xxNB65/status/1507456443385266179

24. NB65, X, March 6, 2022, https://x.com/xxNB65/status/1500420652565737472

25. Sunny, "NB65: Riesen-Blamage mit falschem Kaspersky Source Code-Leak", *Tarnkappe.info*, March 10, 2022, https://tarnkappe.info/artikel/nb65-riesen-blamage-mit-falschem-kaspersky-source-code-leak-221205.html

26. "Mosekspertiza", Distributed Denial of Secrets, accessed May 7, 2024, https://ddosecrets.com/wiki/Mosekspertiza

27. NB65, X, April 3, 2022, https://x.com/xxNB65/status/1510484074070224896

28. NB65, X, April 6, 2022, https://x.com/xxNB65/status/1511472012925050880

29. Later shared the info they obtained with DDoSecrets.

30. NB65, X, April 18, 2022, https://x.com/xxNB65/status/1515860688673165313; NB65, X, May 12, 2022, https://x.com/xxNB65/status/1524823948206145537

31. This entire attack timeline is based on tweets from NB65 and is thus not otherwise verified. There are some reports about NB65's activities, however, all that I could find based their reporting on NB65's Twitter account, not really help-

ing with verification. On encryption mechanism: NB65, X, April 25, 2022, https://x.com/xxNB65/status/1518594916431060992; for later hacks, also see: Waqas, "Anonymous NB65 Claims Hack on Russian Payment Processor Qiwi", HackRead, May 9, 2022, https://www.hackread.com/anonymous-nb65-hacki-russia-payment-processor-qiwi/; NB65, X, May 1, 2022, https://x.com/xxNB65/status/1520663353395486725

32. "Security Vendors' Analysis", *VirusTotal*, accessed May 7, 2024, https://www.virustotal.com/gui/file/7f6dbd9fa0cb7ba2487464c824b6d7e16ace9d4cd15e4452df4c9a9fd6bd1907

33. Lawrence Abrams, "Hackers use Conti's leaked ransomware to attack Russian companies", *Bleeping Computer*, April 9, 2022, https://www.bleepingcomputer.com/news/security/hackers-use-contis-leaked-ransomware-to-attack-russian-companies/

34. As Abrams notes, several research reports suggest that in May 2022, BlackByte was among the ransomware groups targeting Latin American governments. By August 2022, BlackByte had updated their operations, launching a new version that featured a new data leak site and adopted new extortion methods inspired by LockBit. Abrams, "BlackByte Ransomware Gang Is Back with New Extortion Tactics". Also see: Vitali Kremez and Bogulslavskiy, "Hydra with Three Heads: BlackByte & The Future of Ransomware Subsidiary Groups", *Advintel*, May 19, 2022, https://web.archive.org/web/20220519211037/https://www.advintel.io/post/hydra-with-three-heads-blackbyte-the-future-of-ransomware-subsidiary-groups

35. According to Google Threat Analysis Group (TAG), some former members of Conti also repurposed their techniques targeting Ukrainian organizations, the Ukrainian government, and European humanitarian and non-profit organizations. Pierre-Marc Bureau, "Initial access broker repurposing techniques in targeted attacks against Ukraine," September 7, 2022, https://blog.google/threat-analysis-group/initial-access-broker-repurposing-techniques-in-targeted-attacks-against-ukraine/

36. Trend Micro Research, "Ransomware Spotlight: Black Basta", *Trend Micro*, September 1, 2022, https://www.trendmicro.com/vinfo/us/security/news/ransomware-spotlight/ransomware-spotlightblackbasta

37. According to a report, the operators of BlackBasta primarily target organizations in the United States but have attacked other countries. Office of Information Security—Securing One HHS, "HC3: Threat Profile—Threat Profile: Black Basta", Health Sector Cybersecurity Coordination Center, March 15, 2023, https://www.hhs.gov/sites/default/files/black-basta-threat-profile.pdf

38. *Palo Alto Networks Unit 42* and *ZeroFox Intelligence* were quick to note similarities of both group's data leak sites and victim recovery portals but did not directly link Black Basta to Conti.

39. Microsoft Threat Intelligence, "Ransomware as a Service: Understanding the

Cybercrime Gig Economy and How to Protect Yourself", *Microsoft Security Blog*, May 9, 2022, https://www.microsoft.com/en-us/security/blog/2022/05/09/ransomware-as-a-service-understanding-the-cybercrime-gig-economy-and-how-to-protect-yourself/; Blackbasta is also linked to other groups, including FIN7. According to a news report from Bill Toulas: "A backdoor that FIN7 developed in 2018 and still uses was discovered within this EDR. This same backdoor connects to an IP address that FIN7 also uses regularly. Furthermore, additional evidence of a connection between the groups is found in their attack techniques—specifically, the employment of Cobalt Strike". Bill Toulas, "Black Basta ransomware gang linked to the FIN7 hacking group", *Bleeping Computer*, November 3, 2022, https://www.bleepingcomputer.com/news/security/black-basta-ransomware-gang-linked-to-the-fin7-hacking-group/

40. Initial versions of this attack involved deceiving the victim into downloading the BazarLoader malware through documents with malicious macros. Kristopher Russo, "Threat Assessment: Luna Moth Callback Phishing Campaign", *Unit 42*, November 21, 2022, https://unit42.paloaltonetworks.com/luna-moth-call-back-phishing/; also see: Duncan, "BazarCall Method: Call Centers Help Spread BazarLoader Malware".

41. AdvIntel reports that, angered by what they viewed as a "betrayal" and the ensuing success of their former subordinates (Silent Ransom), Conti Team Two, which had rebranded itself as Quantum, started their own callback phishing operations. According to AdvIntel, this new phase, named Jormungandr, began on May 13, 2022. Yelisey Boguslavskiy and Marley Smith, "'BazarCall' Advisory: Essential Guide to Attack Vector that Revolutionized Data Breaches", Internet Archive: WayBackMachine, January 26, 2023, https://web.archive.org/web/20230126213202/https://www.advintel.io/post/bazarcall-advisory-the-essential-guide-to-call-back-phishing-attacks-that-revolutionized-the-data; also discussed here: SafeGuard Cyber Team, "BazarCall: A Double-Edged Attack", *Safeguard Cyber*, September 7, 2022, https://www.safeguardcyber.com/blog/security/3-bazarcall-phishing-groups-wreaking-havoc

42. Eg. Unit 42: "While Unit 42 cannot confirm Silent Ransom's tie to Conti at this time, we are monitoring this closely for attribution. These cases show a clear evolution of tactics that suggests the threat actor is continuing to improve the efficiency of their attack. Cases analyzed at the beginning of the campaign targeted individuals at small- and medium-sized businesses in the legal industry. In contrast, cases later in the campaign indicate a shift in victimology to include individuals at larger targets in the retail sector".

43. Russo, "Threat Assessment: Luna Moth Callback Phishing Campaign".

44. This statement was made by Yelisey Bohuslavskiy. In his research, he also notes that the group is "a gang of former members of Team 2 within the Conti syndicate". Lawrence Abrams, "New Royal Ransomware Emerges in Multi-Million Dollar Attacks", *Bleeping Computer*, September 29, 2022, https://www.bleepingcomputer.com/news/security/new-royal-ransomware-emerges-in-multi-

million-dollar-attacks/; Ionut Ilascu, "Researchers Link 3AM Ransomware to Conti, Royal Cybercrime Gangs", *Bleeping Computer*, January 20, 2024, https://www.bleepingcomputer.com/news/security/researchers-link-3am-ransomware-to-conti-royal-cybercrime-gangs/

45. Abrams, "New Royal Ransomware Emerges in Multi-Million Dollar Attacks"; Cybereason Global SOC & Cybereason Security Research Teams, "Royal Rumble: Analysis of Royal Ransomware".

46. Royal employs the OpenSSL library and the AES256 encryption algorithm, using ReadFile to process targeted files, encrypts the data, and subsequently rewrites the encrypted content using WriteFile and SetFilePointerEx. The file extension is then altered to ".royal" through the API call MoveFileExW. Cybereason states that the adaptability of Royal's encryption approach is probably intended to dodge detection, suggesting a sophisticated development in ransomware tactics akin to those used by Conti. Abrams, "New Royal Ransomware Emerges in Multi-Million Dollar Attacks"; Cybereason Global SOC & Cybereason Security Research Teams, "Royal Rumble: Analysis of Royal Ransomware".

47. Cyber Threat Intelligence Team, "ThreeAM Ransomware", *Intrinsec*, December 2023, https://www.intrinsec.com/wp-content/uploads/2024/01/TLP-CLEAR-2024-01-09-ThreeAM-EN-Information-report.pdf

48. Ionut Ilascu, "Researchers Link 3AM Ransomware to Conti, Royal Cybercrime Gangs".

49. Catalin Cimpanu, "Ryuk Ransomware Crew Makes $640,000 in Recent Activity Surge"; Cyber Signals, "Extortion Economics Ransomware's New Business Model", 2022, https://query.prod.cms.rt.microsoft.com/cms/api/am/binary/RE54L7v; Umawing, "Karakurt Extortion Group: Threat Profile."; Lawrence Abrams, "BlackByte Ransomware Decryptor Released to Recover Files for Free", *Bleeping Computer*, October 19, 2021, https://www.bleepingcomputer.com/news/security/blackbyte-ransomware-decryptor-released-to-recover-files-for-free/; Jim Walter, "From the Front Lines—3 New and Emerging Ransomware Threats Striking Businesses in 2022", *SentinelOne*, June 22, 2022, https://www.sentinelone.com/blog/from-the-front-lines-3-new-and-emerging-ransomware-threats-striking-businesses-in-2022/; Office of Information Security, "HC3: Threat Profile—Threat Profile: Black Basta"; SafeGuard Cyber Team, "BazarCall: A Double-Edged Attack."; Abrams, "BlackByte Ransomware Gang Is Back with New Extortion Tactics"; Jason Reaves and Joshua Platt, "Reverse Engineering", *Walmart Global Tech Blog*, March 12, 2024, https://medium.com/walmartglobaltech/tagged/reverse-engineering; BushidoToken, "The Continuity of Conti"; Cybereason, "Royal Rumble: Analysis of Royal Ransomware"; Lawrence Abrams, "Meet Akira—A New Ransomware Operation Targeting the Enterprise", *Bleeping Computer*, May 7, 2023, https://www.bleepingcomputer.com/news/security/meet-akira-a-new-ransomware-operation-targeting-the-enterprise/; Cyber Threat Intelligence Team, "ThreeAM Ransomware"; Trend Micro Research, "Ransomware

Spotlight: Black Basta."; Oren Biderman et al., "Luna Moth: The Threat Actors Behind Recent False Subscription Scams", *Sygnia*, July 1, 2022, https://www.sygnia.co/blog/luna-moth-false-subscription-scams/

50. Note that this shift in the ransomware ecosystem is not only due to the implosion of Conti.

51. Abrams, "Conti Ransomware Shuts down Operation, Rebrands into Smaller Units".

52. Also see David Uberti, "Russia-Linked Ransomware Groups Are Changing Tactics to Dodge Crackdowns", *The Wall Street Journal*, June 2, 2022, https://www.wsj.com/articles/russia-linked-ransomware-groups-are-changing-tactics-to-dodge-crackdowns-11654178400

CONCLUSION

1. The term "countermeasures" is used here in a general sense to describe approaches aimed at countering ransomware attacks and mitigating their impact. It does not refer to the specific legal meaning of countermeasures in international law, which entails a defined set of actions taken by a state to induce compliance by another state under strict conditions.

2. A longer discussion about this arrest was provided in chapter seven. Kai Biermann, Maria Fedorova, Karsten Polke-Majewski, and Hakan Tanriverdi, "Hacker Gruppe Conti—Don Stern und die Hackermafia", *Die Zeit*, December 11, 2022, https://web.archive.org/web/20230103020739/https://www.zeit.de/digital/2022-12/conti-hackergruppe-russland-ransomsoftware-cyberangriffe

3. Sabrina Weston, "Irish police seize Conti domains used in HSE ransomware attack", *ITPro*, September 6, 2021, https://www.itpro.com/security/ransomware/360786/irish-police-seize-conti-domains-used-in-hse-ransomware-attack

4. Sean Lyngaas, "Biden administration sanctions cryptocurrency exchange in effort to cut off revenue to ransomware groups", *CNN*, September 21, 2021, https://edition.cnn.com/2021/09/21/politics/us-cryptocurrency-sanctions/index.html

5. A darknet marketplace is a type of online store that operates within the hidden sections of the internet known as the dark web. Anthony Blinken, "Sanctions on Darknet Market and Ransomware-Enabling Virtual Currency Exchange", *U.S. Department of State*, April 5, 2022, https://www.state.gov/sanctions-on-darknet-market-and-ransomware-enabling-virtual-currency-exchange/; United States Department of the Treasury, "Treasury Sanctions Russia-Based Hydra, World's Largest Darknet Market, and Ransomware-Enabling Virtual Currency Exchange Garantex", April 5, 2022, https://home.treasury.gov/news/press-releases/jy0701; Ruth Fulterer, "Hydra ist tot: Deutschland sperrt den grössten Darknet-Marktplatz der Welt", *NZZ*, April 5, 2022, https://www.nzz.ch/technologie/deutsche-ermittler-schliessen-hydra-den-groessten-darknet-marktplatz-der-welt-ld.1678073?reduced=true

6. United States Department of the Treasury, "U.S. Treasury Issues First-Ever Sanctions on a Virtual Currency Mixer, Targets DPRK Cyber Threats", May 6, 2022, https://home.treasury.gov/news/press-releases/jy0768; Mitchell Clark, "US punishes Blender.io for helping North Korea launder millions in stolen Axie crypto", *The Verge*, May 7, 2022, https://www.theverge.com/2022/5/6/23060544/us-treasury-blender-io-sanctions-bitcoin-mixer

7. FBI: Federal Bureau of Investigation, "Conti Ransomware Attacks Impact Healthcare and First Responder Networks", May 20, 2021, https://www.aha.org/system/files/media/file/2021/05/fbi-tlp-white-report-conti-ransomware-attacks-impact-healthcare-and-first-responder-networks-5–20–21.pdf

8. Sergiu Gatlan, "FBI, CISA, and NSA warn of escalating Conti ransomware attacks", *Bleeping Computer*, September 22, 2021, https://www.bleepingcomputer.com/news/security/fbi-cisa-and-nsa-warn-of-escalating-conti-ransomware-attacks/

9. Sergiu Gatlan, "Australian govt raises alarm over Conti ransomware attacks", *Bleeping Computer*, December 10, 2021, https://www.bleepingcomputer.com/news/security/australian-govt-raises-alarm-over-conti-ransomware-attacks/

10. Ned Price, "Reward Offers for Information to Bring Conti Ransomware Variant Co-Conspirators to Justice", *U.S. Department of State*, May 6, 2022, https://www.state.gov/reward-offers-for-information-to-bring-conti-ransomware-variant-co-conspirators-to-justice/. Also see: Sergiu Gatlan, "US offers $15 million reward for info on Conti ransomware gang", *Bleeping Computer*, May 7, 2022, https://www.bleepingcomputer.com/news/security/us-offers-15-million-reward-for-info-on-conti-ransomware-gang/

11. UK Government, "UK cracks down on ransomware actors", February 9, 2023, https://www.gov.uk/government/news/uk-cracks-down-on-ransomware-actors; United States Department of the Treasury, "United States and United Kingdom Sanction Members of Russia-Based Trickbot Cybercrime Gang", February 9, 2023, https://home.treasury.gov/news/press-releases/jy1256; Chainalysis Team, "The U.S. and U.K. Sanction Members of Russia-Based Trickbot Cybercrime Gang", *Chainalysis*, February 9, 2023, https://www.chainalysis.com/blog/trickbot-ransomware-sanctions/

12. NCA: National Crime Agency, "Russian ransomware group hit with new sanctions", September 7, 2023, https://web.archive.org/web/20240105184629/https://www.nationalcrimeagency.gov.uk/news/russian-ransomware-group-hit-with-new-sanctions; Lawrence Abrams, "US and UK sanction 11 TrickBot and Conti cybercrime gang members", *Bleeping Computer*, September 7, 2023, https://www.bleepingcomputer.com/news/security/us-and-uk-sanction-11-trickbot-and-conti-cybercrime-gang-members/

13. Some government actions can impact multiple facets of the framework simultaneously. For instance, shutting down a leak site affects both the operational capabilities and the branding efforts of a ransomware group.

14. Organizations have a variety of measures at their disposal to defend against and

recover from ransomware attacks, including automated backups, improved end-point security, regular awareness training, controlled email systems, timely software updates, and strict password policies. However, this conclusion shifts the focus from organizational measures to governmental actions. It examines the roles that governments can play, possibly in partnership with the private sector, to combat ransomware threats. For information on organizational measures, see: Allan Liska, *Ransomware. Understand. Prevent. Recover* (ActualTech Media: 2021); Jon DiMaggio, *The Art of Cyberwarfare: An Investigator's Guide to Espionage, Ransomware, and Organized Cybercrime* (No Starch Press: 2022); Ryan, *Ransomware Revolution*.

15. CISA: Cybersecurity and Infrastructure Security Agency, "New Stop Ransomware.gov website—The U.S. Government's One-Stop Location to Stop Ransomware", July 15, 2021, https://www.cisa.gov/news-events/alerts/2021/07/15/new-stopransomwaregov-website-us-governments-one-stop-location-stop

16. The effectiveness of joint alerts increases when the FBI and CISA provide detailed information on how a ransomware group infiltrates systems. For instance, their August 2022 joint alert on MedusaLocker highlighted that the group primarily exploits Remote Desktop Protocol (RDP) vulnerabilities to penetrate victims' networks, enabling organizations to implement more targeted defenses. For other cases: Cybersecurity & Infrastructure Security Agency, "Conti Ransomware", March 9, 2022, https://www.cisa.gov/news-events/alerts/2021/09/22/conti-ransomware; Cybersecurity & Infrastructure Security Agency, "BlackMatter Ransomware", October 18, 2021, https://www.cisa.gov/news-events/cybersecurity-advisories/aa21–291a; Cybersecurity & Infrastructure Security Agency, #StopRansomware: Play Ransomware, December 18, 2023, https://www.cisa.gov/news-events/cybersecurity-advisories/aa23–352a; Cybersecurity and Infrastructure Security Agency, "FBI and CISA Release Advisory on Scattered Spider Group", November 16, 2023, https://www.cisa.gov/news-events/alerts/2023/11/16/fbi-and-cisa-release-advisory-scattered-spider-group; Cybersecurity & Infrastructure Security Agency, #StopRansomware: Black Basta, May 10, 2024, https://www.cisa.gov/news-events/cybersecurity-advisories/aa24–131a; Cybersecurity and Infrastructure Security Agency, #StopRansomware: Daixin Team, October 26, 2022, https://www.cisa.gov/news-events/cybersecurity-advisories/aa22–294a

17. Additionally, around $8 million ransomware proceeds from groups like Conti, REvil, and Ryuk were transferred on the market. United States Department of the Treasury, "Treasury Sanctions Russia-Based Hydra, World's Largest Darknet Market, and Ransomware-Enabling Virtual Currency Exchange Garantex", April 5, 2022, https://home.treasury.gov/news/press-releases/jy0701

18. Fulterer, "Hydra ist tot: Deutschland sperrt den grössten Darknet-Marktplatz der Welt".

19. Sergiu Gatlan, "FBI seizes BreachForums after arresting its owner Pompompurin

in March", *Bleeping Computer*, June 23, 2023, https://www.bleepingcomputer.
com/news/security/fbi-seizes-breachforums-after-arresting-its-owner-pom-
pompurin-in-march/

20. For a general overview of botnet takedowns, see: Jason Healey, Neil Jenkins,
    and JD Work, "Defenders Disrupting Adversaries: Framework, Dataset, and
    Case Studies of Disruptive Counter-Cyber Operations", *12th International
    Conference on Cyber Conflict. 20/20 Vision: The Next Decade*, edited by T. Jančárková,
    L. Lindström, M. Signoretti, I. Tolga, and G. Visky, NATO CCDCOE
    Publications, 2020, https://ccdcoe.org/uploads/2020/05/CyCon_2020_14_
    Healey_Jenkins_Work.pdf

21. The other key botnet takedown related to Ryuk and Conti was Emotet in
    January 2022, as discussed in chapter two. Also see: United States Department
    of Justice, "Emotet Botnet Disrupted in International Cyber Operation", January
    28, 2021, https://www.justice.gov/opa/pr/emotet-botnet-disrupted-inter-
    national-cyber-operation; Biermann, Fedorova, Polke-Majewski, and Tanriverdi,
    "Hackergruppe Conti—Don Stern und die Hackermafia".

22. Kurt Baker, "What is Trickbot Malware?", *CrowdStrike*, October 23, 2023,
    https://www.crowdstrike.com/cybersecurity-101/malware/trickbot/

23. Krebs, "Attacks Aimed at Disrupting the Trickbot Botnet", October 2, 2020,
    https://krebsonsecurity.com/2020/10/attacks-aimed-at-disrupting-the-trick-
    bot-botnet/

24. Ellen Nakashima, "Cyber Command has sought to disrupt the world's largest
    botnet hoping to reduce its potential impact on the election", *The Washington
    Post*, October 9, 2020, https://www.washingtonpost.com/national-security/
    cyber-command-trickbot-disrupt/2020/10/09/19587aae-0a32-11eb-a166-
    dc429b380d10_story.html. For the US Cyber Command vision on persistent
    engagement, see: U.S. Cyber Command, "Achieve and Maintain Cyberspace
    Superiority: Command Vision for US Cyber Command", 2018, https://www.
    cybercom.mil/Portals/56/Documents/USCYBERCOM%20Vision%20
    April%202018.pdf. For more on persistent engagement theory, see: Michael
    Fischerkeller, Emily Goldman, Richard Harknett, *Cyber Persistence Theory:
    Redefining National Security in Cyberspace* (Oxford University Press: 2022).

25. It has been suggested that the goal of these interventions was not to permanently
    dismantle Trickbot, but rather to temporarily disrupt the operators' activities,
    preventing them from creating chaos during the US Presidential elections.
    Nakashima, "Cyber Command has sought to disrupt the world's largest botnet
    hoping to reduce its potential impact on the election"; Ionut Ilascu, "TrickBot
    botnet targeted in takedown operations, little impact seen", *Bleeping Computer*,
    October 12, 2020, https://www.bleepingcomputer.com/news/security/
    trickbot-botnet-targeted-in-takedown-operations-little-impact-seen/.   On
    Microsoft's actions, see: Tom Burt, "New action to combat ransomware ahead
    of U.S. elections", October 12, 2020, https://blogs.microsoft.com/on-the-
    issues/2020/10/12/trickbot-ransomware-cyberthreat-us-elections/

26. There have been instances where the government, aware of encryption vulnerabilities, has shared this information publicly to aid recovery efforts for organizations targeted by ransomware. For example, in March 2021, the FBI issued an alert about a vulnerability in the encryption process used by Mamba ransomware. First detected in 2016, Mamba, also known as HDDCryptor, was significant for using open-source software to encrypt entire storage volumes instead of just individual files. Activity for Mamba ransomware ramped up with a new variant emerging in late 2019. Although it did not operate through an affiliate program, Mamba was considered one of the top threats at the time. For more on the FBI alert of Mamba: FBI: Federal Bureau of Investigation, "FBI Flash: Mamba Ransomware", *StopRansomware.org*, https://www.cisa.gov/stopransomware/fbi-flash-mamba-ransomware; Ionut Ilascu, "FBI exposes weakness in Mamba ransomware, DiskCryptor", *Bleeping Computer*, March 26, 2021, https://www.bleepingcomputer.com/news/security/fbi-exposes-weakness-in-mamba-ransomware-diskcryptor/. EUROPOL European Cybercrime Centre also occasionally shares decryptors as part of the 'No More Ransom' project: e.g. EUROPOL, "No More Ransom update: Belgian Federal Police releases free decryption keys for the Cryakl ransomware," (2024), https://www.europol.europa.eu/cms/sites/default/files/documents/ransomware_infographic.pdf

27. The Chainalysis team, "How the Dutch National Police Tricked Prolific Ransomware Strain Deadbolt Into Giving Up Victim Decryption Keys", *Chainalysis*, March 1, 2023, https://www.chainalysis.com/blog/deadbolt-ransomware-strain-tricked-into-giving-up-decryption-keys/

28. Protos Staff, "Dutch police recover 90% of victim decryption keys in ransomware scam", *Protos*, March 1, 2023, https://protos.com/dutch-police-recover-90-of-victim-decryption-keys-in-ransomware-scam/

29. Also see the No More Ransom portal, launched in 2016: https://www.nomoreransom.org/. EUROPOL, "Hit by ransomware? No More Ransom now offers 136 free tools to rescue your files", July 26, 2022, https://www.europol.europa.eu/media-press/newsroom/news/hit-ransomware-no-more-ransom-now-offers-136-free-tools-to-rescue-your-files; Chainalysis, "How the Dutch National Police Tricked Prolific Ransomware Strain Deadbolt Into Giving Up Victim Decryption Keys", *Chainalysis*, March 1, 2023, https://www.chainalysis.com/blog/deadbolt-ransomware-strain-tricked-into-giving-up-decryption-keys/. Note that a lot of times the decrypter is provided by a private sector company. See, for example, Bitfender releases of GandCrab decryption to let victims who got infected by the ransomware recover files without paying the ransom demand: Catalin Cimpanu, "Bitdefender releases third GandCrab ransomware free decrypter in the past year", *ZDNET*, February 19, 2019, https://www.zdnet.com/article/rupert-goodwins-diary-4010004362/

30. U.S. Department of Justice, Office of Public Affairs, "U.S. Department of Justice Disrupts Hive Ransomware Variant", January 26, 2023, https://www.justice.gov/opa/pr/us-department-justice-disrupts-hive-ransomware-variant

31. Kaseya asked New Zealand-based security firm Emsisoft to create a fresh decryption tool, which Kaseya released the following day. However, for some victims, the assistance came too late. Ellen Nakashima and Rachel Lerman, "FBI held back ransomware decryption key from businesses to run operation targeting hackers", September 21, 2021, https://www.washingtonpost.com/national-security/ransomware-fbi-revil-decryption-key/2021/09/21/4a9417d0-f15f-11eb-a452–4da5fe48582d_story.html

32. For longer discussion on Kaseya and Revil, see chapters one and eight.

33. In 2023, all 50 member nations of the International Counter Ransomware Initiative committed to a policy statement agreeing not to meet ransom demands from cybercriminals. Jessica Lyons, "Formal ban on ransomware payments? Asking orgs nicely to not cough up ain't working", *The Register*, January 3, 2024, https://www.theregister.com/2024/01/03/ban_ransomware_payments/. For a comparative legal analysis, see: Sean O'Connell, "To Ban Ransomware Payments or Not to Ban Ransomware Payments: The Problems Drafting Legislation in Response to Ransomware", *Journal of International Business and Law,* 22(1), 2023, https://scholarlycommons.law.hofstra.edu/jibl/vol22/iss1/6

34. NCSC: National Cyber Security Centre, "Encryption malware—What next?", January 5, 2022, https://www.ncsc.admin.ch/ncsc/en/home/infos-fuer/infos-unternehmen/vorfall-was-nun/ransomware.html. The Swiss Criminal Code does not categorically classify paying a ransom as a criminal offense. Yan Borboën, "Ransomware as a business model—Legal aspects of ransom payment", *PwC*, https://www.pwc.ch/en/insights/cybersecurity/ransom-payment.html#legal-aspect

35. Ciaran Martin, "Cyber ransoms are too profitable. Let's make paying illegal", *The Times*, March 4, 2024, https://www.thetimes.com/article/cyber-ransoms-are-too-profitable-lets-make-paying-illegal-kc8cmhxs0. According to Laurie Mercer from HackerOne, "Enforcing a ransomware payment ban is like banning smoking—you know it's good for society in the long run but in the short term, it is difficult to stop getting a quick fix". Dan Raywood, "Proposals to Ban Ransomware Payments Rumoured", *SC Media UK*, May 22, 2024, https://insight.scmagazineuk.com/proposals-to-ban-ransomware-payments-rumoured

36. Emisoft Malware Lab, "The State of Ransomware in the U.S.: Report and Statistics 2023", January 2, 2024, https://www.emsisoft.com/en/blog/44987/the-state-of-ransomware-in-the-u-s-report-and-statistics-2023/

37. Also see: O'Connell, "To Ban Ransomware Payments or Not to Ban Ransomware Payments: The Problems Drafting Legislation in Response to Ransomware".

38. A variation of this argument on country-wide bans was also made in the IST report, arguing that if a ban is not introduced internationally at the same time, the ransomware groups will just shift their focus to other sectors/countries and make the situation worse for them. Ransomware Task Force, "Combating Ransomware: A Comprehensive Framework for Action: Key Recommendations from the Ransomware Task Force", *IST: Institute for Security and Technology,*

2021, https://securityandtechnology.org/wp-content/uploads/2021/09/IST-Ransomware-Task-Force-Report.pdf

39. BBC News, "Liberty Reserve digital money service forced offline", May 27, 2013, https://www.bbc.com/news/technology-22680297

40. Ibid.

41. Brian Krebs, "Reports: Liberty Reserve Founder Arrested, Site Shuttered", *Krebs on Security*, May 25, 2013, https://krebsonsecurity.com/2013/05/reports-liberty-reserve-founder-arrested-site-shuttered/

42. Catalin Cimpanu, "Reveton ransomware distributor sentenced to six years in prison in the UK", *ZDNET*, April 9, 2019, https://www.zdnet.com/article/reveton-ransomware-distributor-sentenced-to-six-years-in-prison-in-the-uk/

43. This included by ransomware groups like Zeppelin, SunCrypt, Mamba, Dharma, Lockbit. U.S. Department of Justice, "Justice Department Investigation Leads to Takedown of Darknet Cryptocurrency Mixer that Processed Over $3 Billion of Unlawful Transactions", March 15, 2023, https://www.justice.gov/usao-edpa/pr/justice-department-investigation-leads-takedown-darknet-cryptocurrency-mixer-processed

44. James Reddick, "'Prolific' crypto money laundering platform ChipMixer shuttered by Germany, US", *The Record*, March 15, 2023, https://therecord.media/chipmixer-takedown-cryptocurrency-money-laundering-europol-doj; BKA: Bundes Kriminal Amt, "BKA schaltet weltweit größten Geldwäschedienst im Darknet ab", March 15, 2023, https://www.bka.de/DE/Presse/Listenseite_Pressemitteilungen/2023/Presse2023/230314_Geldwaesche_Darknet.html

45. U.S. Department of Justice, "Justice Department Investigation Leads to Takedown of Darknet Cryptocurrency Mixer that Processed Over $3 Billion of Unlawful Transactions".

46. A concerning trend is emerging where young hackers from Western countries are collaborating with Russian criminal networks to carry out highly disruptive ransomware attacks. Notably, a group mainly consisting of native English speakers, known as Scattered Spider, has targeted over 130 organizations. Yet, their residence in Western countries tends to facilitate arrests. In fact, in June 2024, a 22-year-old British man, purportedly the leader of Scattered Spider, was arrested in Spain as he attempted to board a flight to Italy. Bill Whitaker, Aliza Chasan, Graham Messick, Jack Weingart, "Criminal exploits of Scattered Spider earn respect of Russian ransomware hackers", April 14, 2024, https://www.cbsnews.com/news/scattered-spider-blackcat-hackers-ransomware-team-up-60-minutes/; *Murcia Today*, "VIDEO: FBI take down UK hacker in Spain for stealing 27M USD of Bitcoins", June 14, 2024, https://murciatoday.com/video-fbi-take-down-uk-hacker-in-spain-for-stealing-27m-usd-of-bitcoins_1000077536-a.html#:~:text=A%2022-year-old%20British,board%20a%20flight%20to%20Italy; Brian Krebs, "Alleged Boss of 'Scattered Spider' Hacking Group Arrested", June 15, 2024, https://krebsonsecurity.com/2024/06/alleged-boss-of-scattered-spider-hacking-group-arrested/

47. Lawrence Abrams, "US govt will pay you $10 million for info on Conti ransomware members", *Bleeping Computer*, August 11, 2022, https://www.bleepingcomputer.com/news/security/us-govt-will-pay-you-10-million-for-info-on-conti-ransomware-members/

48. The White House, "Joint Statement of the Ministers and Representatives from the Counter Ransomware Initiative Meeting October 2021," October 2021, https://www.whitehouse.gov/briefing-room/statements-releases/2021/10/14/joint-statement-of-the-ministers-and-representatives-from-the-counter-ransomware-initiative-meeting-october-2021/

49. The White House, "FACT SHEET: The Second International Counter Ransomware Initiative Summit," November 1, 2022, https://www.whitehouse.gov/briefing-room/statements-releases/2022/11/01/fact-sheet-the-second-international-counter-ransomware-initiative-summit/

50. The White House, "International Counter Ransomware Initiative 2023 Joint Statement," November 2023, https://www.whitehouse.gov/briefing-room/statements-releases/2023/11/01/international-counter-ransomware-initiative-2023-joint-statement/

51. Institute for Security and Technology, "Ransomware Task Force (RTF)," https://securityandtechnology.org/ransomwaretaskforce/

52. Ransomware Task Force, "Combating Ransomware A Comprehensive Framework for Action: Key Recommendations from the Ransomware Task Force."; Ransomware Task Force, "Ransomware Task Force: Doubling Down," *IST: Institute for Security and Policy*, 2024, https://securityandtechnology.org/virtual-library/reports/ransomware-task-force-doubling-down/

53. The cornerstone report from the RTF outlines four key goals; deterring ransomware attacks, disrupting the ransomware business model, helping organizations prepare for ransomware attacks, and enhancing responses to such attacks. These goals lead to detailed objectives, which in turn guide specific recommended actions.

54. There are other aspects, for which I could not find evidence that they happened. But not unlikely. For example, governments can remove traces of ransomware groups from the internet. This includes deleting archives, forum posts, and any digital footprint that could lend credibility or historical significance to the groups.

55. See the press conference; also see the BCC story and Alexander Martin, "LockBit held victims' data even after receiving ransom payments to delete it", *The Record*, February 21, 2024, https://therecord.media/lockbit-lied-about-deleting-exfiltrated-data-after-ransom-payments; for more information on the NCA led operation, see: NCA: National Crime Agency, "International investigation disrupts the world's most harmful cyber crime group", February 20, 2024, https://www.nationalcrimeagency.gov.uk/news/nca-leads-international-investigation-targeting-worlds-most-harmful-ransomware-group; Matt Burgess, "A Global Police Operation Just Took Down the Notorious LockBit Ransomware Gang",

Wired, February 20, 2024, https://www.wired.com/story/lockbit-ransomware-takedown-website-nca-fbi/; Alexander Martin, "Police plan week of LockBit revelations after capturing 'unprecedented' intelligence from gang's infrastructure", *The Record*, February 20, 2024, https://therecord.media/lockbit-ransomware-gang-shutdown-cybercrime-intelligence-captured

56. Remarks at "The Oxford Cyber Forum," the European Cyber Conflict Research Initiative and Blavatnik School of Government, University of Oxford, June 27, 2024.

57. Although this remains most actively done by the private sector instead of governments.

58. Sophos X-ops, "Press and Pressure".

59. Coverage can unintentionally encourage further attacks and increase pressure on victims. We have observed this dynamic in the case of Conti's ransom demand to Graff; initially, Graff refused to pay the ransom. However, several months later, when the DailyMail and other media outlets began publishing personal details from the leak, Graff succumbed to the pressure and paid. This manipulation of media pressure is a tactic well-understood by ransomware groups, harking back to the earliest cases of double extortion. As I noted in chapter one, when the Maze ransomware group sought to increase pressure on its victims to pay, they approached Lawrence Abrams of Bleeping Computer, hoping he would publish details of their activities and escalate the urgency for their targets.

60. Olga Belogolova, Lee Foster, Thomas Rid, and Gavin Wilde, "Don't Hype the Disinformation Threat," *Foreign Affairs*, May 3, 2024, https://www.foreignaffairs.com/russian-federation/dont-hype-disinformation-threat

61. Cherilyn Ireton and Julie Posetti (eds.), *Journalism, fake news & disinformation: handbook for journalism education and training*, UNESCO, 2018, https://unesdoc.unesco.org/ark:/48223/pf0000265552

62. Whitney Phillips, "The Oxygen of Amplification: Better Practices for Reporting," *Data & Society*, 2018, https://datasociety.net/library/oxygen-of-amplification/

63. Eg. Jean Paul Marthoz and Khalid Aoutail, "Media and the coverage of terrorism: manual for trainers and journalism educators", UNESCO, 2022, https://unesdoc.unesco.org/ark:/48223/pf0000380356

64. The LinkedIn post from Greenberg continues to state: "… Now RansomHub has sent us samples of patient records and a contract that appear to have been taken from Change Healthcare. If RansomHub really does possess Change Healthcare's stolen data—and two ransomware researchers who reviewed the samples now believe it does—then this represents a new kind of worst case scenario for a ransomware victim: pay an eight-figure ransom and still face a second extortion demand from the hackers' jilted partner". Andy Greenberg, Linkedin, April 2024, https://www.linkedin.com/posts/andygreenbergjournalist_change-healthcare-faces-another-ransomware-activity-7184620359

134912512-kVEK/. The article: Andy Greenberg and Matt Burgess, "Change Healthcare Faces Another Ransomware Threat—and It Looks Credible", Wired, April 12, 2024, https://www.wired.com/story/change-healthcare-ransom-hub-threat/

65. The downside is that this type of reporting can also lead to a lot of self-advertisement of cybersecurity companies too. See for example Darktrace's report on how its technology stopped the Akira ransomware from harming some of their clients. Darktrace, "How Darktrace Stopped Akira Ransomware", September 13, 2023, https://darktrace.com/blog/akira-ransomware-how-darktrace-foiled-another-novel-ransomware-attack

66. Also see the chapter for Russia's rare instance of collaboration regarding the arrest of some REvil affiliates.

67. Eg. the Summit on Police Ransomware already in 2011, coordinated by Europol. EUROPOL, "Europol hosts expert meeting to combat the spread of "Police Ransomware"", May 7, 2012, https://www.europol.europa.eu/media-press/newsroom/news/europol-hosts-expert-meeting-to-combat-spread-of-police-ransomware. Or Operation GoldDust/Quicksand: INTERPOL, "Joint global ransomware operation sees arrests and criminal network dismantled", November 8, 2021, https://www.interpol.int/News-and-Events/News/2021/Joint-global-ransomware-operation-sees-arrests-and-criminal-network-dismantled; Sergiu Gatlan, "REvil ransomware affiliates arrested in Romania and Kuwait", *Bleeping Computer*, November 8, 2021, https://www.bleepingcomputer.com/news/security/revil-ransomware-affiliates-arrested-in-romania-and-kuwait/; EUROPOL, "Five affiliates to Sodinokibi/REvil unplugged", November 8, 2021, https://www.europol.europa.eu/media-press/newsroom/news/five-affiliates-to-sodinokibi/revil-unplugged; Operation Cronos 2024; Lockbit Takedown EUROPOL, "Law enforcement disrupt world's biggest ransomware operation", February 20, 2024, https://www.europol.europa.eu/media-press/newsroom/news/law-enforcement-disrupt-worlds-biggest-ransomware-operation

68. The White House, "Joint Statement of the Ministers and Representatives from the Counter Ransomware Initiative Meeting October 2021", October 14, 2021, https://www.whitehouse.gov/briefing-room/statements-releases/2021/10/14/joint-statement-of-the-ministers-and-representatives-from-the-counter-ransomware-initiative-meeting-october-2021/; Australian Department of Home Affairs, "Global task force to fight ransomware commences operations", January 23, 2023, https://www.homeaffairs.gov.au/news-media/archive/article?itemId=1013

SELECTED BIBLIOGRAPHY

Abrams, Lawrence. "Allied Universal Breached by Maze Ransomware, Stolen Data Leaked". *Bleeping Computer*, November 21, 2019. https://www.bleepingcomputer.com/news/security/allied-universal-breached-by-maze-ransomware-stolen-data-leaked/.

————. "Another Ransomware Will Now Publish Victims' Data If Not Paid". *Bleeping Computer*, December 12, 2019. https://www.bleepingcomputer.com/news/security/another-ransomware-will-now-publish-victims-data-if-not-paid/.

————. "Asteelflash Electronics Maker Hit by REvil Ransomware Attack". *Bleeping Computer*, April 2, 2021. https://www.bleepingcomputer.com/news/security/asteelflash-electronics-maker-hit-by-revil-ransomware-attack/.

————. "Babuk Ransomware's Full Source Code Leaked on Hacker Forum". *Bleeping Computer*, September 3, 2021. https://www.bleepingcomputer.com/news/security/babuk-ransomwares-full-source-code-leaked-on-hacker-forum/.

————. "BazarCall Malware Uses Malicious Call Centers to Infect Victims". *Bleeping Computer*, March 31, 2021. https://www.bleepingcomputer.com/news/security/bazarcall-malware-uses-malicious-call-centers-to-infect-victims/.

————. "BlackByte Ransomware Decryptor Released to Recover Files for Free". *Bleeping Computer*, October 19, 2021. https://www.bleepingcomputer.com/news/security/blackbyte-ransomware-decryptor-released-to-recover-files-for-free/.

————. "BlackByte Ransomware Gang Is Back with New Extortion Tactics". *Bleeping Computer*, August 17, 2022. https://www.bleepingcomputer.com/news/security/blackbyte-ransomware-gang-is-back-with-new-extortion-tactics/.

————. "Conti Ransomware Shows Signs of Being Ryuk's Successor". *Bleeping Computer*, July 9, 2020. https://www.bleepingcomputer.com/news/security/conti-ransomware-shows-signs-of-being-ryuks-successor/.

————. "Conti Ransomware Shuts down Operation, Rebrands into Smaller Units". *Bleeping Computer*, May 19, 2022. https://www.bleepingcomputer.com/news/security/conti-ransomware-shuts-down-operation-rebrands-into-smaller-units/.

————. "Conti Ransomware Source Code Leaked by Ukrainian Researcher". *Bleeping Computer*, March 1, 2022. https://www.bleepingcomputer.com/news/security/conti-ransomware-source-code-leaked-by-ukrainian-researcher/.

————. "FBI Links Diavol Ransomware to the TrickBot Cybercrime Group". *Bleeping Computer*, January 20, 2022. https://www.bleepingcomputer.com/news/security/fbi-links-diavol-ransomware-to-the-trickbot-cybercrime-group/.

————. "Hackers use Conti's leaked ransomware to attack Russian companies". *Bleeping Computer*, April 9, 2022. https://www.bleepingcomputer.com/news/security/hackers-use-contis-leaked-ransomware-to-attack-russian-companies/.

————. "JBS Paid $11 Million to REvil Ransomware, $22.5M First Demanded". *Bleeping Computer*, June 10, 2021. https://www.bleepingcomputer.com/news/security/jbs-paid-11-million-to-revil-ransomware-225m-first-demanded/.

————. "Leaked Babuk Locker Ransomware Builder Used in New Attacks". *Bleeping Computer*, June 30, 2021. https://www.bleepingcomputer.com/news/security/leaked-babuk-locker-ransomware-builder-used-in-new-attacks/.

————. "Meet Akira—A New Ransomware Operation Targeting the Enterprise". *Bleeping Computer*, May 7, 2023. https://www.bleepingcomputer.com/news/security/meet-akira-a-new-ransomware-operation-targeting-the-enterprise/.

————. "More Conti Ransomware Source Code Leaked on Twitter out of Revenge". *Bleeping Computer*, March 20, 2022. https://www.bleepingcomputer.com/news/security/more-conti-ransomware-source-code-leaked-on-twitter-out-of-revenge/.

————. "New Royal Ransomware Emerges in Multi-Million Dollar Attacks". *Bleeping Computer*, September 29, 2022. https://www.bleepingcomputer.com/news/security/new-royal-ransomware-emerges-in-multi-million-dollar-attacks/.

————. "Popular Russian Hacking Forum XSS Bans All Ransomware Topics". *Bleeping Computer*, May 13, 2021. https://www.bleepingcomputer.com/news/security/popular-russian-hacking-forum-xss-bans-all-ransomware-topics/.

————. "REvil Ransomware Gang's Web Sites Mysteriously Shut Down". *Bleeping Computer*, July 13, 2021. https://www.bleepingcomputer.com/news/security/revil-ransomware-gangs-web-sites-mysteriously-shut-down/.

————. "REvil Ransomware Is Back in Full Attack Mode and Leaking Data". *Bleeping Computer*, September 11, 2021. https://www.bleepingcomputer.com/news/security/revil-ransomware-is-back-in-full-attack-mode-and-leaking-data/.

————. "REvil Ransomware Shuts down Again after Tor Sites Were Hijacked". *Bleeping Computer*, October 17, 2021. https://www.bleepingcomputer.com/news/security/revil-ransomware-shuts-down-again-after-tor-sites-were-hijacked/.

————. "REvil Ransomware's Servers Mysteriously Come Back Online". *Bleeping Computer*, September 7, 2021. https://www.bleepingcomputer.com/news/security/revil-ransomwares-servers-mysteriously-come-back-online/.

————. "Ryuk Ransomware Likely Behind New Orleans Cyberattack". *Bleeping*

Computer, December 15, 2019. https://www.bleepingcomputer.com/news/security/ryuk-ransomware-likely-behind-new-orleans-cyberattack/.

————. "Ryuk Ransomware Partners with TrickBot to Gain Access to Infected Networks". *Bleeping Computer*, January 12, 2019. https://www.bleepingcomputer.com/news/security/ryuk-ransomware-partners-with-trickbot-to-gain-access-to-infected-networks/.

————. "Sodinokibi Ransomware Says Travelex Will Pay, One Way or Another". *Bleeping Computer*, January 9, 2020. https://www.bleepingcomputer.com/news/security/sodinokibi-ransomware-says-travelex-will-pay-one-way-or-another/.

————. "Stampado Ransomware Campaign Decrypted before It Started". *Bleeping Computer*, July 22, 2016. https://www.bleepingcomputer.com/news/security/stampado-ransomware-campaign-decrypted-before-it-started/.

————. "The Shark Ransomware Project Allows You to Create Your Own Customized Ransomware". *Bleeping Computer*, August 15, 2016. https://www.bleepingcomputer.com/news/security/the-shark-ransomware-project-allows-to-create-your-own-customized-ransomware/.

————. "US and UK sanction 11 TrickBot and Conti cybercrime gang members". *Bleeping Computer*. September 7, 2023. https://www.bleepingcomputer.com/news/security/us-and-uk-sanction-11-trickbot-and-conti-cybercrime-gang-members/

————. "US govt will pay you $10 million for info on Conti ransomware members". *Bleeping Computer*. August 11, 2022. https://www.bleepingcomputer.com/news/security/us-govt-will-pay-you-10-million-for-info-on-conti-ransomware-members/

AdvIntel. "Persist, Brick, Profit—TrickBot Offers New 'TrickBoot' UEFI-Focused Functionality". December 3, 2020. https://web.archive.org/web/20220416031255/https://www.advintel.io/post/persist-brick-profit-trickbot-offers-new-trickboot-uefi-focused-functionality.

————. "Digital 'Pharmacusa' II: The 'GandCrab' Phenomenon". Internet Archive: WayBackMachine, January 26, 2023. https://web.archive.org/web/20230126230909/https://www.advintel.io/post/digital-pharmacusa-ii-the-gandcrab-phenomenon

AFP. "'Shocking' Hack of Psychotherapy Records in Finland Affects Thousands". *The Guardian*, October 26, 2020. sec. World News. https://www.theguardian.com/world/2020/oct/26/tens-of-thousands-psychotherapy-records-hacked-in-finland.

AIDS 88 Summary: A Practical Synopsis of the IV International Conference, Stockholm, Sweden. NCJRS Virtual Library, 1988. https://www.ojp.gov/ncjrs/virtual-library/abstracts/aids-88-summary-practical-synopsis-iv-international-conference

Amit Malik, Vikas Taneja, and Sameer Patil. "Analysis of Shadow Brokers Release". Cysinfo. April, 2017. https://cysinfo.com/wp-content/uploads/2017/04/Shadow_release_updated.pdf.

SELECTED BIBLIOGRAPHY

Ansoff, Igor H. *Corporate Strategy*. McGraw-Hill, New York, NY, 1965.

Antoniuk, Daryna. "Cyber Community Mourns Renowned Researcher Vitali Kremez". *The Record*, November 3, 2022. https://therecord.media/cyber-community-mourns-renowned-researcher-vitali-kremez.

Aodha, Gráinne Ní. "'Real Arms Race' on Defending Irish Health System against Cyber Attacks". *BreakingNews.ie*, February 9, 2023. https://www.breakingnews.ie/ireland/real-arms-race-on-defending-irish-health-system-against-cyber-attacks-1430272.html.

Appelbaum, Jacob, Aaron Gibson, Claudio Guarnieri, Andy Müller-Maguhn, Laura Poitras, Marcel Rosenbach, Leif Ryge, Hilmar Schmundt, and Michael Sontheimer. "New Snowden Docs Indicate Scope of NSA Preparations for Cyber Battle". *Der Spiegel*, January 17, 2015. sec. International. https://www.spiegel.de/international/world/new-snowden-docs-indicate-scope-of-nsa-preparations-for-cyber-battle-a-1013409.html.

Australian Department of Home Affairs. "Global task force to fight ransomware commences operations". January 23, 2023. https://www.homeaffairs.gov.au/news-media/archive/article?itemId=1013.

Baker, Kurt. "History of Ransomware". *CrowdStrike*, October 10, 2022. https://www.crowdstrike.com/cybersecurity-101/ransomware/history-of-ransomware/.

———. "What Is TrickBot Malware?" *CrowdStrike*, October 3, 2023. https://www.crowdstrike.com/cybersecurity-101/malware/trickbot/.

Baydakova, Anna. "Ransomware Gang Extorted 725 BTC in One Attack, On-Chain Sleuths Find". CoinDesk: *Consensus Magazine*, May 17, 2022. https://www.coindesk.com/layer2/2022/05/17/ransomware-gang-extorted-725-btc-in-one-attack-on-chain-sleuths-find/.

BBC News. "REvil Ransomware Gang Arrested in Russia". January 14, 2022. https://www.bbc.com/news/technology-59998925.

Beauchamp, Zack. "The WikiLeaks-Russia Connection Started Way before the 2016 Election". *Vox*, January 6, 2017. https://www.vox.com/world/2017/1/6/14179240/wikileaks-russia-ties.

Bellingcat Investigation Team. "Guccifer Rising? Months-Long Phishing Campaign on ProtonMail Targets Dozens of Russia-Focused Journalists and NGOs". *Bellingcat*, August 10, 2019. https://www.bellingcat.com/news/uk-and-europe/2019/08/10/guccifer-rising-months-long-phishing-campaign-on-protonmail-targets-dozens-of-russia-focused-journalists-and-ngos/.

Bevilacqua, Betsy. "The Law Is Finally Catching Up With Ransomware Criminals". *Wired*, February 21, 2022. https://www.wired.com/story/law-fighting-ransomware-criminals/.

Biderman, Oren, Tomer Lahiyani, Noam Lifshitz, and Ori Porag. "Luna Moth: The Threat Actors Behind Recent False Subscription Scams". *Sygnia*, July 1, 2022. https://www.sygnia.co/blog/luna-moth-false-subscription-scams/.

Biermann, Kai, Maria Fedorova, Karsten Polke-Majewski, and Hakan Tanriverdi.

SELECTED BIBLIOGRAPHY

"Hacker Gruppe Conti—Don Stern und die Hackerfirma". *Die Zeit*. December 11, 2022. https://web.archive.org/web/20230103020739/https://www.zeit.de/digital/2022–12/conti-hackergruppe-russland-ransomsoftware-cyberangriffe.

Binary Defense. "New Ransomware 'Diavol' Being Dropped by Trickbot". April 18, 2023. https://www.binarydefense.com/resources/threat-watch/new-ransomware-diavol-being-dropped-by-trickbot/.

Bleiweiss, Arianne. "New Russian-Speaking Forum—A New Place for RaaS?" *KELA Cyber Threat Intelligence*, July 28, 2021. https://www.kelacyber.com/new-russian-speaking-forum-a-new-place-for-raas/.

Blinken, Anthony. "Sanctions on Darknet Market and Ransomware-Enabling Virtual Currency Exchange". *U.S. Department of State*. April 5, 2022. https://www.state.gov/sanctions-on-darknet-market-and-ransomware-enabling-virtual-currency-exchange/.

Bogusalvskiy, Yelisey, and Vitali Kremez. "DisCONTInued: The End of Conti's Brand Marks New Chapter For Cybercrime Landscape". Internet Archive: WayBackMachine, October 26, 2022. https://web.archive.org/web/20221026025639/https://www.advintel.io/post/discontinued-the-end-of-conti-s-brand-marks-new-chapter-for-cybercrime-landscape.

Bogusalvskiy, Yelisey. "The TrickBot Saga's Finale Has Aired: Spinoff is Already in the Works". Internet Archive: WayBackMachine, March 1, 2022. https://web.archive.org/web/20220301022043/https://www.advintel.io/post/the-trickbot-saga-s-finale-has-aired-but-a-spinoff-is-already-in-the-works.

Boguslavskiy, Yelisey, and Marley Smith. "'BazarCall' Advisory: Essential Guide to Attack Vector that Revolutionized Data Breaches". Internet Archive: WayBackMachine, January 26, 2023. https://web.archive.org/web/20230126213202/https://www.advintel.io/post/bazarcall-advisory-the-essential-guide-to-call-back-phishing-attacks-that-revolutionized-the-data.

Boom, Daniel Van. "Forget Bitcoin: Inside the Insane World of Altcoin Cryptocurrency Trading". CNET, April 13, 2021. https://www.cnet.com/personal-finance/crypto/features/beyond-bitcoin-the-wild-world-of-altcoin-cryptocurrency-trading/.

Broadcom. "Several New Ransomware Variants Use the Leaked Conti Source Code". December 22, 2022. https://www.broadcom.com/support/security-center/protection-bulletin/several-new-ransomware-variants-use-the-leaked-conti-source-code.

Buchanan, Ben. *The Cybersecurity Dilemma: Hacking, Trust and Fear Between Nations*. Oxford University Press, 2017.

Burgess, Matt. "A Global Police Operation Just Took Down the Notorious LockBit Ransomware Gang". *Wired*. February 20, 2024. https://www.wired.com/story/lockbit-ransomware-takedown-website-nca-fbi/.

———. "Leaked Ransomware Docs Show Conti Helping Putin From the Shadows". *Wired*, March 18, 2022. https://www.wired.com/story/conti-ransomware-russia/.

———. "The Big, Baffling Crypto Dreams of a $180 Million Ransomware Gang". *Wired*, March 17, 2022. https://www.wired.com/story/conti-ransomware-crypto-payments/.

———. "Unmasking Trickbot, One of the World's Top Cybercrime Gangs". *Wired*, August 30, 2023. https://www.wired.com/story/trickbot-trickleaks-bentley/.

Burt, Tom. "New action to combat ransomware ahead of U.S. elections". October 12, 2020. https://blogs.microsoft.com/on-the-issues/2020/10/12/trickbot-ransomware-cyberthreat-us-elections/

BushidoToken Threat Intel. "Lessons from the Conti Leaks". April 17, 2022. https://blog.bushidotoken.net/2022/04/lessons-from-conti-leaks.html.

———. "The Continuity of Conti". November 17, 2022. https://blog.bushidotoken.net/2022/11/the-continuity-of-conti.html.

Canadian Centre for Cyber Security. "National Cyber Threat Assessment 2023–2024". October 28, 2022. https://www.cyber.gc.ca/en/guidance/national-cyber-threat-assessment-2023-2024.

Catalin Cimpanu, *Extortion Economics Ransomware's New Business Model: Cyber Signals*. Microsoft, 2022. https://query.prod.cms.rt.microsoft.com/cms/api/am/binary/RE54L7v

Chainalysis Team. "How the Dutch National Police Tricked Prolific Ransomware Strain Deadbolt Into Giving Up Victim Decryption Keys". *Chainalysis*. March 1, 2023. https://www.chainalysis.com/blog/deadbolt-ransomware-strain-tricked-into-giving-up-decryption-keys/.

———. "The U.S. and U.K. Sanction Members of Russia-Based Trickbot Cybercrime Gang". *Chainalysis*. February 9, 2023. https://www.chainalysis.com/blog/trickbot-ransomware-sanctions/.

———. "U.S. and U.K. Sanction 11 Members of Trickbot Ransomware Group". *Chainalysis*, September 7, 2023. https://www.chainalysis.com/blog/trickbot-ransomware-malware-sanctions-september-2023/.

Check Point Research Team. "CPR Reveals Leaks of Conti Ransomware Group". *Check Point Blog*, March 11, 2022. https://blog.checkpoint.com/security/check-point-research-revels-leaks-of-conti-ransomware-group/.

Chesney, Robert and Max Smeets, eds. *Deter, Disrupt, or Deceive: Assessing Cyber Conflict as an Intelligence Contest*. Georgetown University Press, 2023.

Chiu, Richard. "Crime Gang Apologises to Graff Jewellers over Data Leak". *Jeweller*, November 8, 2021. https://www.jewellermagazine.com/Article/10138/Crime-gang-apologises-to-Graff-Jewellers-over-data-leak.

Cimpanu, Catalin. "Bitdefender releases third GandCrab ransomware free decrypter in the past year". *ZDNET*. February 19, 2019. https://www.zdnet.com/article/rupert-goodwins-diary-4010004362/.

———. "North Korean Hackers Used Hermes Ransomware to Hide Recent Bank Heist". *Bleeping Computer*, October 17, 2017. https://www.bleepingcomputer.com/news/security/north-korean-hackers-used-hermes-ransomware-to-hide-recent-bank-heist/.

————. "Reveton ransomware distributor sentenced to six years in prison in the UK". *ZDNET*. April 9, 2019. https://www.zdnet.com/article/reveton-ransomware-distributor-sentenced-to-six-years-in-prison-in-the-uk/.

————. "Ryuk Ransomware Crew Makes $640,000 in Recent Activity Surge". *Bleeping Computer*, August 21, 2018. https://www.bleepingcomputer.com/news/security/ryuk-ransomware-crew-makes-640–000-in-recent-activity-surge/.

CISA: Cybersecurity and Infrastructure Security Agency. "Karakurt Data Extortion Group". December 12, 2023. https://www.cisa.gov/news-events/cybersecurity-advisories/aa22–152a.

————. "Ransomware 101". Stopransomware.org. https://www.cisa.gov/stopransomware/ransomware-101.

Clark, Mitchell. "US punishes Blender.io for helping North Korea launder millions in stolen Axie crypto". *The Verge*. May 7, 2022. https://www.theverge.com/2022/5/6/23060544/us-treasury-blender-io-sanctions-bitcoin-mixer.

Colin Cowie. "Yanlouwang Ransomware Leaks". *Colins Security Blog*. https://www.th3protocol.com/2022/Yanlouwang-Leaks.

Constantin, Lucian. "REvil Ransomware Explained: A Widespread Extortion Operation". CSO Online, November 12, 2021. https://www.csoonline.com/article/570101/revil-ransomware-explained-a-widespread-extortion-operation.html.

PWC. "Conti Cyber Attack on the HSE Independent Post: Incident Review". 2021. https://www.hse.ie/eng/services/publications/conti-cyber-attack-on-the-hse-full-report.pdf.

Counter Threat Unit Research Team. "GOLD ULRICK Continues Conti Operations despite Public Disclosures". Secureworks, April 21, 2022. https://www.secureworks.com/blog/gold-ulrick-continues-conti-operations-despite-public-disclosures.

Coveware. "Conti Ransomware Recovery, Payment & Decryption Statistics". Accessed May 7, 2024. https://www.coveware.com/conti-ransomware.

Coveware: Ransomware Recovery First Responders. "Q3 Ransomware Demands Rise: Maze Sunsets & Ryuk Returns". November 4, 2020. https://www.coveware.com/blog/q3–2020-ransomware-marketplace-report.

Cox, Joseph. "Hackers Breach Russian Space Research Institute Website". *Vice*, March 3, 2022. https://www.vice.com/en/article/z3n8ea/hackers-breach-russian-space-research-institute-website.

CPR. "Leaks of Conti Ransomware Group Paint Picture of a Surprisingly Normal Tech Start-Up… Sort Of". *Check Point Research*, March 10, 2022. https://research.checkpoint.com/2022/leaks-of-conti-ransomware-group-paint-picture-of-a-surprisingly-normal-tech-start-up-sort-of/.

CrowdStrike Intel Team. "Wizard Spider Update: Resilient, Reactive and Resolute". *CrowdStrike*, October 16, 2020. https://www.crowdstrike.com/blog/wizard-spider-adversary-update/.

SELECTED BIBLIOGRAPHY

Cryptome. "The Alternative History of Public-Key Cryptography". Accessed May 4, 2024. https://cryptome.org/ukpk-alt.htm.

Crystal Investigations Team. "The Conti Leaks Part One". *Crystal Intelligence*, March 31, 2022. https://crystalintelligence.com/investigations/the-conti-leaks-part-one/.

Cyber Threat Intelligence Team. "ThreeAM Ransomware". Intrinsec, December 2023. https://www.intrinsec.com/wp-content/uploads/2024/01/TLP-CLEAR-2024-01-09-ThreeAM-EN-Information-report.pdf.

Cybereason. "Royal Rumble: Analysis of Royal Ransomware". Accessed May 5, 2024. https://www.cybereason.com/blog/royal-ransomware-analysis.

Cybereason Nocturnus Team. "Dropping Anchor: From a TrickBot Infection to the Discovery of the Anchor Malware". *Cybereason*. Accessed May 6, 2024. https://www.cybereason.com/blog/research/dropping-anchor-from-a-trickbot-infection-to-the-discovery-of-the-anchor-malware.

CyberTalk. "Conti Ransomware Gang Shutdown, Conti Ransomware Rebranding 2022". May 20, 2022. https://www.cybertalk.org/2022/05/20/conti-ransomware-gang-shuts-down-rebranding-into-smaller-units/.

Cyble. "New Ransomware Strains Emerging From Leaked Conti's Source Code". December 22, 2022. https://cyble.com/blog/new-ransomware-strains-emerging-from-leaked-contis-source-code/.

CYJAX. "Who Is Trickbot? Analysis of the Trickbot Leaks". July 2022. https://www.cyjax.com/wp-content/uploads/2022/07/Who-is-Trickbot.pdf.

DiMaggio, Jon. "A Behind the Scenes Look into Investigating Conti Leaks". *Analyst1*, March 21, 2022. https://analyst1.com/a-behind-the-scenes-look-into-investigating-conti-leaks/.

———. "A History of REvil". *Anaylst1*. Accessed May 6, 2024. https://analyst1.com/history-of-revil/.

———. "Ransomware Diaries: Volume 1". *Analyst1*, January 16, 2023. https://analyst1.com/ransomware-diaries-volume-1/.

Doffman, Zak. "Russia Linked To Cyberattacks On Bellingcat Researchers Investigating GRU (Updated)". *Forbes*, July 26, 2019. https://www.forbes.com/sites/zakdoffman/2019/07/26/russian-intelligence-cyberattacked-journalists-hacking-encrypted-email-accounts/.

Dong, Chuong. "Conti Ransomware". December 15, 2020. https://cdong1012.github.io//reverse%20engineering/2020/12/15/ContiRansomware/.

Dudley, Renee, and Daniel Golden. *The Ransomware Hunting Team: A Band of Misfits' Improbable Crusade to Save the World from Cybercrime*. Farrar, Straus and Giroux, 2022.

Duncan, Brad. "BazarCall Method: Call Centers Help Spread BazarLoader Malware". *Unit 42*, May 19, 2021. https://unit42.paloaltonetworks.com/bazarloader-malware/.

Elliptic. "Troubled Dark Web Carding Market Loses Another Key Vendor as FBI Seizes SSNDOB". June 8, 2022. https://www.elliptic.co/blog/troubled-dark-web-carding-market-loses-another-key-vendor-as-fbi-seizes-ssndob.

SELECTED BIBLIOGRAPHY

Emisoft Malware Lab. "The State of Ransomware in the U.S.: Report and Statistics 2023". January 2, 2024. https://www.emsisoft.com/en/blog/44987/the-state-of-ransomware-in-the-u-s-report-and-statistics-2023/.

eSentire. "Analysis of Leaked Conti Intrusion Procedures by eSentire's Threat Response Unit (TRU)". March 18, 2022. https://www.esentire.com/blog/analysis-of-leaked-conti-intrusion-procedures-by-esentires-threat-response-unit-tru.

Eugenio, Dexter. "A Targeted Campaign Break-Down—Ryuk Ransomware". Check Point Research, August 20, 2018. https://research.checkpoint.com/2018/ryuk-ransomware-targeted-campaign-break/.

Europol. "Internet Organised Crime Threat Assessment (IOCTA) 2020". December 7, 2021. Accessed May 4, 2024. https://www.europol.europa.eu/publications-events/main-reports/internet-organised-crime-threat-assessment-iocta-2020.

FBI: Federal Bureau of Investigation. "Conti Ransomware Attacks Impact Healthcare and First Responder Networks". May 20, 2021. https://www.aha.org/system/files/media/file/2021/05/fbi-tlp-white-report-conti-ransomware-attacks-impact-healthcare-and-first-responder-networks-5-20-21.pdf.

Ferbrache, David. *A Pathology of Computer Viruses*. Springer Science & Business Media, 1992.

Figueroa, Marco, Napoleon Bing, And Bernard Silvestrini. "The Conti Leaks Insight into a Ransomware Unicorn". Internet Archive: WayBackMachine, December 7, 2023. https://web.archive.org/web/20231207134956/https://www.breach-quest.com/blog/conti-leaks-insight-into-a-ransomware-unicorn/.

FileZilla Wiki. "FileZilla FTP Server—FileZilla Wiki". Accessed May 6, 2024. https://wiki.filezilla-project.org/FileZilla_FTP_Server.

Flashpoint. "How Ransomware Has Become an 'Ethical' Dilemma in the Eastern European Underground". September 20, 2017. https://flashpoint.io/blog/ransomware-ethical-dilemma-eastern-european-underground/.

Fokker, John, and Jambul Tologonov. "Conti Leaks: Examining the Panama Papers of Ransomware". *Trellix*, March 31, 2022. https://www.trellix.com/en-gb/blogs/research/conti-leaks-examining-the-panama-papers-of-ransomware/.

Fokker, John. "Dismantling a Prolific Cybercriminal Empire: REvil Arrests and Reemergence". *Trellix*, September 29, 2022. https://www.trellix.com/blogs/research/dismantling-a-prolific-cybercriminal-empire/.

FSB of Russia. "Illegal Activities of Members of the Organized Criminal Community Were Suppressed". January 14, 2022. http://www.fsb.ru/fsb/press/message/single.htm%21id%3D10439388%40fsbMessage.html.

Gartzke, Erik. "The Myth of Cyberwar: Bringing War in Cyberspace Back Down to Earth". *International Security* 38, no. 2 (October 2013): 41–73. https://doi.org/10.1162/ISEC_a_00136.

Gatlan, Sergiu. "FBI seizes BreachForums after arresting its owner Pompompurin in March". *Bleeping Computer*. June 23, 2023. https://www.bleepingcomputer.com/news/security/fbi-seizes-breachforums-after-arresting-its-owner-pompom-purin-in-march/.

SELECTED BIBLIOGRAPHY

————. "FBI, CISA, and NSA warn of escalating Conti ransomware attacks". *Bleeping Computer*. September 22, 2021. https://www.bleepingcomputer.com/news/security/fbi-cisa-and-nsa-warn-of-escalating-conti-ransomware-attacks/.

————. "REvil Ransomware Affiliates Arrested in Romania and Kuwait". *Bleeping Computer*, November 8, 2021. https://www.bleepingcomputer.com/news/security/revil-ransomware-affiliates-arrested-in-romania-and-kuwait/.

————. "REvil ransomware affiliates arrested in Romania and Kuwait". *Bleeping Computer*. November 8, 2021. https://www.bleepingcomputer.com/news/security/revil-ransomware-affiliates-arrested-in-romania-and-kuwait/.

Glenny, Misha. *Darkmarket: How Hackers Became the New Mafia*. Vintage, 2012.

Granger, Diana. "Fatboy Ransomware-as-a-Service Emerges on Russian-Language Forum". *Recorded Future*, May 4, 2017. https://www.recordedfuture.com/blog/fatboy-ransomware-analysis.

Gray, Ian W., Jack Cable, Benjamin Brown, Vlad Cuiujuclu, and Damon McCoy. "Money Over Morals: A Business Analysis of Conti Ransomware". arXiv, April 23, 2023. https://doi.org/10.48550/arXiv.2304.11681.

Greenberg, Andy, and Matt Burgess. "Change Healthcare Faces Another Ransomware Threat—and It Looks Credible". *Wired*. April 12, 2024. https://www.wired.com/story/change-healthcare-ransomhub-threat/.

Greenberg, Andy. "The Ransomware Hackers Made Some Real Amateur Mistakes". *Wired*, May 15, 2017. https://www.wired.com/2017/05/wannacry-ransomware-hackers-made-real-amateur-mistakes/.

————. *Sandworm: A New Era of Cyberwar and the Hunt for the Kremlin's Most Dangerous Hackers*. Doubleday, 2019.

Greig, Jonathan. "FBI Decision to Withhold Kaseya Ransomware Decryption Keys Stirs Debate". *ZDNET*, September 24, 2021. https://www.zdnet.com/article/fbi-decision-to-withhold-kaseya-ransomware-decryption-keys-stirs-debate/.

————. "REvil Ransomware Operators Claim Group Is Ending Activity Again, Victim Leak Blog Now Offline". *ZDNET*, October 18, 2021. https://www.zdnet.com/article/revil-ransomware-operators-claim-group-is-ending-activity-again-happy-blog-now-offline/.

Greig, Jonathan. "Several Colombian Government Ministries Hampered by Ransomware Attack". The Record, September 15, 2023. https://therecord.media/colombia-government-ministries-cyberattack.

Hammond, Charlotte, and Chris Caridi. "Analysis of Diavol Ransomware Reveals Possible Link to TrickBot Gang". *Security Intelligence*, August 17, 2021. https://securityintelligence.com/posts/analysis-of-diavol-ransomware-link-trickbot-gang/.

Hammond, Charlotte, and Ole Villadsen. "The Trickbot/Conti Crypters: Where Are They Now?" *Security Intelligence*, June 27, 2023. https://securityintelligence.com/x-force/trickbot-conti-crypters-where-are-they-now/.

————. "Trickbot Group's AnchorDNS Backdoor Upgrades to AnchorMail". *Security Intelligence*, February 25, 2022. https://securityintelligence.com/posts/new-malware-trickbot-anchordns-backdoor-upgrades-anchormail/.

SELECTED BIBLIOGRAPHY

Hat, Black. *Keynote: Black Hat at 25: Where Do We Go from Here?* Youtube, November 18, 2022. https://www.youtube.com/watch?v=doRZwCbbyNs.

Healey, Jason, Neil Jenkins, and JD Work. "Defenders Disrupting Adversaries: Framework, Dataset, and Case Studies of Disruptive Counter-Cyber Operations". *12th International Conference on Cyber Conflict. 20/20 Vision: The Next Decade.* edited by T. Jančárková, L. Lindström, M. Signoretti, I. Tolga, and G. Visky, NATO CCDCOE Publications. 2020. https://ccdcoe.org/uploads/2020/05/CyCon_2020_14_Healey_Jenkins_Work.pdf.

——. "The Spectrum of National Responsibility for Cyberattacks". *The Brown Journal of World Affairs* 18, no. 1 (2011): 57–70. https://www.jstor.org/stable/24590776.

HSE NQPSD. "A Mixed Methods Analysis of the Effectiveness of the Patient Safety Risk Mitigation Strategies Following a Healthcare ICT Failure". 2022. http://hdl.handle.net/10147/631586.

HSE.ie. "If You Received a Letter from the HSE about the Cyber-Attack". Accessed May 5, 2024. https://www2.hse.ie/services/cyber-attack/received-letter/.

Huntress. "Persistence in Cybersecurity". Accessed May 5, 2024. https://www.huntress.com/defenders-handbooks/persistence-in-cybersecurity.

Ilascu, Ionut. "Diavol Ransomware Sample Shows Stronger Connection to TrickBot Gang". *Bleeping Computer*, August 18, 2021. https://www.bleepingcomputer.com/news/security/diavol-ransomware-sample-shows-stronger-connection-to-trickbot-gang/.

——. "FBI exposes weakness in Mamba ransomware, DiskCryptor". *Bleeping Computer*. March 26, 2021. https://www.bleepingcomputer.com/news/security/fbi-exposes-weakness-in-mamba-ransomware-diskcryptor/.

——. "Hackers Ask for $5.3 Million Ransom, Turn Down $400k, Get Nothing". *Bleeping Computer*, September 5, 2019. https://www.bleepingcomputer.com/news/security/hackers-ask-for-53-million-ransom-turn-down-400k-get-nothing/.

——. "How Conti Ransomware Hacked and Encrypted the Costa Rican Government". *Bleeping Computer*, July 21, 2022. https://www.bleepingcomputer.com/news/security/how-conti-ransomware-hacked-and-encrypted-the-costa-rican-government/.

——. "How Ryuk Ransomware Operators Made $34 Million from One Victim". *Bleeping Computer*, November 7, 2020. https://www.bleepingcomputer.com/news/security/how-ryuk-ransomware-operators-made-34-million-from-one-victim/.

——. "Karakurt Revealed as Data Extortion Arm of Conti Cybercrime Syndicate". *Bleeping Computer*, April 15, 2022. https://www.bleepingcomputer.com/news/security/karakurt-revealed-as-data-extortion-arm-of-conti-cybercrime-syndicate/.

——. "Ransomware Threat Surge, Ryuk Attacks about 20 Orgs per Week". *Bleeping Computer*, October 6, 2020. https://www.bleepingcomputer.com/

news/security/ransomware-threat-surge-ryuk-attacks-about-20-orgs-per-week/.

————. "Researchers Link 3AM Ransomware to Conti, Royal Cybercrime Gangs". *Bleeping Computer*, January 20, 2024. https://www.bleepingcomputer.com/news/security/researchers-link-3am-ransomware-to-conti-royal-cybercrime-gangs/.

————. "REvil Ransomware Found Buyer for Trump Data, Now Targeting Madonna". *Bleeping Computer*, May 18, 2020. https://www.bleepingcomputer.com/news/security/revil-ransomware-found-buyer-for-trump-data-now-targeting-madonna/.

————. "Russia Arrests REvil Ransomware Gang Members, Seize $6.6 Million". *Bleeping Computer*, January 14, 2022. https://www.bleepingcomputer.com/news/security/russia-arrests-revil-ransomware-gang-members-seize-66-million/.

————. "Sodinokibi Ransomware Hits Travelex, Demands $3 Million". *Bleeping Computer*, January 6, 2020. https://www.bleepingcomputer.com/news/security/sodinokibi-ransomware-hits-travelex-demands-3-million/.

————. "Sodinokibi, Ryuk Ransomware Drive up Average Ransom to $111,000". *Bleeping Computer*, May 2, 2020. https://www.bleepingcomputer.com/news/security/sodinokibi-ryuk-ransomware-drive-up-average-ransom-to-111-000/.

————. "Translated Conti Ransomware Playbook Gives Insight into Attacks". *Bleeping Computer*, September 2, 2021. https://www.bleepingcomputer.com/news/security/translated-conti-ransomware-playbook-gives-insight-into-attacks/.

————. "TrickBot botnet targeted in takedown operations, little impact seen". *Bleeping Computer*. October 12, 2020. https://www.bleepingcomputer.com/news/security/trickbot-botnet-targeted-in-takedown-operations-little-impact-seen/.

————. "US Accounts for More than Half of World's Ransomware Attacks". *Bleeping Computer*, August 8, 2019. https://www.bleepingcomputer.com/news/security/us-accounts-for-more-than-half-of-worlds-ransomware-attacks/.

Imano, Shunichi, and Fred Gutierrez. "Ransomware Roundup—New Vohuk, ScareCrow, and AERST Variants". *FortiGuard Labs Threat Research*. Fortinet, December 8, 2022. https://www.fortinet.com/blog/threat-research/ransomware-roundup-new-vohuk-scarecrow-and-aerst-variants.

INSIKT Group. "China's PLA Unit 61419 Purchasing Foreign Antivirus Products, Likely for Exploitation". *Record Future*, May 5, 2021. https://www.recordedfuture.com/blog/china-pla-unit-purchasing-antivirus-exploitation.

————. "Latin American Governments Targeted By Ransomware". Recorded Future, June 14, 2022. https://www.recordedfuture.com/blog/latin-american-governments-targeted-by-ransomware.

Intel471. "Understanding the Relationship between Emotet, Ryuk and TrickBot," April 14, 2020. https://intel471.com/blog/understanding-the-relationship-between-emotet-ryuk-and-trickbot.

SELECTED BIBLIOGRAPHY

INTERPOL. "Joint global ransomware operation sees arrests and criminal network dismantled". November 8, 2021. https://www.interpol.int/News-and-Events/News/2021/Joint-global-ransomware-operation-sees-arrests-and-criminal-network-dismantled.

Irwin, Luke. "South Staffordshire Water Targeted by Cyber Attack". *IT Governance UK Blog*, August 16, 2022. https://www.itgovernance.co.uk/blog/south-staffordshire-water-targeted-by-cyber-attack.

Isaac, Anna, Caitlin Ostroff, and Bradley Hope. "Travelex Paid Hackers Multimillion-Dollar Ransom Before Hitting New Obstacles". *The Wall Street Journal*, April 9, 2020. https://web.archive.org/web/20221212140922/https://www.wsj.com/articles/travelex-paid-hackers-multimillion-dollar-ransom-before-hitting-new-obstacles-11586440800.

Ivanov, Anton, and Orkhan Mamedov. "ExPetr/Petya/NotPetya Is a Wiper, Not Ransomware". *SecureList*, June 28, 2017. https://securelist.com/expetrpetya-notpetya-is-a-wiper-not-ransomware/78902/.

James, William, and Steve Scherer. "Russia Trying to Steal COVID-19 Vaccine Data, Say UK, U.S. and Canada". *Reuters*, July 16, 2020, sec. Technology. https://www.reuters.com/article/idUSKCN24H232/.

Janofsky, Adam. "Ransomware Victims Paid More than $600 Million to Cybercriminals in 2021". *The Record*, February 10, 2022. https://therecord.media/ransomware-victims-paid-more-than-600-million-to-cybercriminals-in-2021.

Jansen, Pieter. "Cyberhelden—Episodes". *Cyberhelden.nl*, April 14, 2022. https://www.cyberhelden.nl/episodes/.

Johansmeyer, Tom. "Debunking NotPetya's Cyber Catastrophe Myth". *Binding Hook*, April 10, 2024. https://bindinghook.com/articles-binding-edge/debunking-notpetyas-cyber-catastrophe-myth/.

Junio, Timothy J. "How Probable Is Cyber War? Bringing IR Theory Back In to the Cyber Conflict Debate". *Journal of Strategic Studies* 36, no. 1 (February 2013): 125–33. https://doi.org/10.1080/01402390.2012.739561.

Kapur, Daksh. "Evolution of BazarCall Social Engineering Tactics". Trellix, October 6, 2022. https://www.trellix.com/blogs/research/evolution-of-bazarcall-social-engineering-tactics/.

KELA Targeted Cyber Intelligence. "KELA Intelligence Report: Analysis of Leaked Conti's Internal Data". March 15, 2022. https://ke-la.com/wp-content/uploads/2022/03/KELA-Intelligence-Report-ContiLeaks-1.pdf.

Krebs, Chris, Keynote: Black Hat at 25: Where Do We Go from Here? 2022. https://www.youtube.com/watch?v=doRZwCbbyNs.

KnowBe4. "CryptoWall Ransomware". Accessed May 5, 2024. https://www.knowbe4.com/cryptowall.

———. "Jigsaw Ransomware". Accessed May 6, 2024. https://www.knowbe4.com/jigsaw-ransomware.

———. "Reveton Worm Ransomware". Accessed May 4, 2024. https://www.knowbe4.com/reveton-worm.

SELECTED BIBLIOGRAPHY

Kostka, Cary. "What Is Archiveus Trojan? A Part of the History of Modern Ransomware". Ransomware.org, February 23, 2022. https://ransomware.org/blog/archiveus-trojan-a-part-of-the-history-of-modern-ransomware/.

Kovacs, Nicolas. "Data Analysis of the Shadow Brokers Leak". *Digital Security*. April 16, 2017. https://www.digital.security/en/blog/data-analysis-shadow-brokers-leak.

Krebs, Brian. "Attacks Aimed at Disrupting the Trickbot Botnet". *Krebs on Security*. October 2, 2020. https://krebsonsecurity.com/2020/10/attacks-aimed-at-disrupting-the-trickbot-botnet/.

———. "Conti Ransomware Group Diaries, Part I: Evasion". *Krebs on Security*. March 1, 2022. https://krebsonsecurity.com/2022/03/conti-ransomware-group-diaries-part-i-evasion/.

———. "Conti Ransomware Group Diaries, Part II: The Office". *Krebs on Security*. March 2, 2022. https://krebsonsecurity.com/2022/03/conti-ransomware-group-diaries-part-ii-the-office/.

———. "Conti Ransomware Group Diaries, Part III: Weaponry". *Krebs on Security*. March 4, 2022. https://krebsonsecurity.com/2022/03/conti-ransomware-group-diaries-part-iii-weaponry/.

———. "Conti Ransomware Group Diaries, Part IV: Cryptocrime". *Krebs on Security*. March 8, 2022. https://krebsonsecurity.com/2022/03/conti-ransomware-group-diaries-part-iv-cryptocrime/.

———. "How Does One Get Hired by a Top Cybercrime Gang?" *Krebs on Security*. June 15, 2021. https://krebsonsecurity.com/2021/06/how-does-one-get-hired-by-a-top-cybercrime-gang/.

———. "Is 'REvil' the New GandCrab Ransomware?" *Krebs on Security*. July 15, 2019. https://krebsonsecurity.com/2019/07/is-revil-the-new-gandcrab-ransomware/.

———. "Reports: Liberty Reserve Founder Arrested, Site Shuttered". *Krebs on Security*. May 25, 2013. https://krebsonsecurity.com/2013/05/reports-liberty-reserve-founder-arrested-site-shuttered/.

———. "Who's Behind the GandCrab Ransomware?" *Krebs on Security*. July 8, 2019. https://krebsonsecurity.com/2019/07/whos-behind-the-gandcrab-ransomware/.

Kremez, Vitali, and Yelisey Boguslavskiy. "Backup 'Removal' Solutions—From Conti Ransomware with Love". Internet Archive: WayBackMachine, February 8, 2023. https://web.archive.org/web/20230208191330/https://www.advintel.io/post/backup-removal-solutions-from-conti-ransomware-with-love.

———. "Hydra with Three Heads: BlackByte & The Future of Ransomware Subsidiary Groups". Internet Archive: WayBackMachine, May 19, 2022. https://web.archive.org/web/20220519211037/https://www.advintel.io/post/hydra-with-three-heads-blackbyte-the-future-of-ransomware-subsidiary-groups.

Kremez, Vitali, Yelisey Boguslavskiy, and Marley Smith. "Anatomy of Attack: Truth behind the Costa Rica Government Ransomware 5-Day Intrusion". Internet

SELECTED BIBLIOGRAPHY

Archive: WayBackMachine, November 26, 2022. https://web.archive.org/web/20221126092151/https://www.advintel.io/post/anatomy-of-attack-truth-behind-the-costa-rica-government-ransomware-5-day-intrusion.

Lakshmanan, Ravie. "TrickBot Malware Gang Upgrades Its AnchorDNS Backdoor to AnchorMail". The Hacker News, March 1, 2022. https://thehackernews.com/2022/03/trickbot-malware-gang-upgrades-its.html.

Lee, Micah. "Leaked Chats Show Russian Ransomware Gang Discussing Putin's Invasion of Ukraine". The Intercept, March 14, 2022. https://theintercept.com/2022/03/14/russia-ukraine-conti-russian-hackers/.

Libicki, Martin C. "Cyberspace is Not a Warfighting Domain". *I/S A Journal of Law and Policy for the Information Society* 8, no. 2 (2012): 325–40. http://moritzlaw.osu.edu/students/groups/is/files/2012/02/4.Libicki.pdf.

Libicki, Martin. *Cyberwar and Cyberdeterrence*. Santa Monica: RAND Corporation, 2009.

Liff, Adam P. "Cyberwar: A New 'Absolute Weapon'? The Proliferation of Cyberwarfare Capabilities and Interstate War". *Journal of Strategic Studies* 35, no. 3 (June 2012): 401–28. https://doi.org/10.1080/01402390.2012.663252.

———. "The Proliferation of Cyberwarfare Capabilities and Interstate War, Redux: Liff Responds to Junio". *Journal of Strategic Studies* 36, no. 1 (February 2013): 134–38. https://doi.org/10.1080/01402390.2012.733312.

Liska, Allan. "Is Double/Triple/Whatever Extortion Working?" Ransomware, August 1, 2021. https://ransomwaresommelier.com/p/is-doubletriplewhatever-extortion.

———. "The Etymology of Ransomware". *Ransomware*, June 11, 2023. https://ransomwaresommelier.com/p/the-etymology-of-ransomware.

Lockheed Martin. "*Gaining the Advantage: Applying Cyber Kill Chain Methodology to Network Defense*". 2015. https://www.lockheedmartin.com/content/dam/lockheed-martin/rms/documents/cyber/Gaining_the_Advantage_Cyber_Kill_Chain.pdf.

Logan, Magno, Erika Mendoza, Ryan Maglaque, and Nikko Tamaña. "The State of Ransomware: 2020's Catch-22—Security News". *Trend Micro IE*, February 3, 2021. https://www.trendmicro.com/vinfo/ie/security/news/cybercrime-and-digital-threats/the-state-of-ransomware-2020-s-catch-22.

Luce, Tristan Puech, Laurenne-Sya. "Ransomware: Inside the Former CONTI Group". *RiskInsight*, July 1, 2022. https://www.riskinsight-wavestone.com/en/2022/07/ransomware-inside-the-former-conti-group/.

Lusthaus, Jonathan. *Industry of Anonymity: Inside the Business of Cybercrime*. Harvard University Press, 2018.

———. "Trust in the World of Cybercrime". *Global Crime* 13, 2 (May 2012): 71–94. https://doi.org/10.1080/17440572.2012.674183.

Lyngaas, Sean. "'I Can Fight with a Keyboard': How One Ukrainian IT Specialist Exposed a Notorious Russian Ransomware Gang". *CNN*, March 30, 2022. https://www.cnn.com/2022/03/30/politics/ukraine-hack-russian-ransomware-gang/index.html.

————. "Biden administration sanctions cryptocurrency exchange in effort to cut off revenue to ransomware groups". *CNN*. September 21, 2021. https://edition.cnn.com/2021/09/21/politics/us-cryptocurrency-sanctions/index.html.

Mackenzie, Peter, and Tilly Travers. "What to Expect When You've Been Hit with Conti Ransomware". *Sophos News*, February 16, 2021. https://news.sophos.com/en-us/2021/02/16/what-to-expect-when-youve-been-hit-with-conti-ransomware/.

Marget, Adam. "Ransomware-as-a-Service (RaaS): What It Is & How It Works". *Unitrends*, August 5, 2022. https://www.unitrends.com/blog/ransomware-as-a-service-raas.

Martin, Alexander. "LockBit held victims' data even after receiving ransom payments to delete it". *The Record*. February 21, 2024. https://therecord.media/lockbit-lied-about-deleting-exfiltrated-data-after-ransom-payments.

————. "Police plan week of LockBit revelations after capturing 'unprecedented' intelligence from gang's infrastructure". *The Record*. February 20, 2024. https://therecord.media/lockbit-ransomware-gang-shutdown-cybercrime-intelligence-captured.

————. "Ransomware Gang Posts Breast Cancer Patients' Clinical Photographs". *The Record*, March 6, 2023. https://therecord.media/ransomware-lehigh-valley-alphv-black-cat.

————. "Ransomware Incidents Now Make up Majority of British Government's Crisis Management 'Cobra' Meetings". *The Record*, November 18, 2022. https://therecord.media/ransomware-incidents-now-make-up-majority-of-british-governments-crisis-management-cobra-meetings.

Martinez, Fernando. "REvil's New Linux Version". *AT&T Cybersecurity*. Cybersecurity, July 1, 2021. https://cybersecurity.att.com/blogs/labs-research/revils-new-linux-version.

Matishak, Martin. "U.S. Convenes 30 Countries on Ransomware Threat—without Russia or China". October 13, 2021. https://therecord.media/u-s-convenes-30-countries-on-ransomware-threat-without-russia-or-china.

McAfee Labs. "McAfee ATR Analyzes Sodinokibi Aka REvil Ransomware-as-a-Service—What The Code Tells Us". *McAfee Blog*, October 2, 2019. https://www.mcafee.com/blogs/other-blogs/mcafee-labs/mcafee-atr-analyzes-sodinokibi-aka-revil-ransomware-as-a-service-what-the-code-tells-us/.

McGee, Marianne Kolbasuk. "Breast Cancer Patients Sue Over Breached Exam Photos, Data". *Bank Info Security*, March 14, 2023. https://www.bankinfosecurity.com/breast-cancer-patients-sue-over-breached-exam-photos-data-a-21431.

Mendrez, Rodel, and Lloyd Macrohon. "BlackByte Ransomware—Pt. 1 In-Depth Analysis". *Spiderlabs Blog. Trustwave*, October 15, 2021. https://www.trustwave.com/en-us/resources/blogs/spiderlabs-blog/blackbyte-ransomware-pt-1-in-depth-analysis/.

Microsoft Security Response Center. "Microsoft MSHTML Remote Code Execution Vulnerability: CVE-2021–40444 Security Vulnerability". September 7, 2021.

SELECTED BIBLIOGRAPHY

Accessed May 5, 2024. https://msrc.microsoft.com/update-guide/vulnerability/CVE-2021–40444.

Microsoft Threat Intelligence. "Analyzing Attacks That Exploit the CVE-2021–40444 MSHTML Vulnerability". *Microsoft Security Blog*, September 15, 2021. https://www.microsoft.com/en-us/security/blog/2021/09/15/analyzing-attacks-that-exploit-the-mshtml-cve-2021–40444-vulnerability/.

———. "BazaCall: Phony Call Centers Lead to Exfiltration and Ransomware". *Microsoft Security Blog*, July 29, 2021. https://www.microsoft.com/en-us/security/blog/2021/07/29/bazacall-phony-call-centers-lead-to-exfiltration-and-ransomware/.

———. "Iran Surges Cyber-Enabled Influence Operations in Support of Hamas". Microsoft, February 26, 2024. https://www.microsoft.com/en-us/security/security-insider/intelligence-reports/iran-surges-cyber-enabled-influence-operations-in-support-of-hamas.

———. "New 'Prestige' Ransomware Impacts Organizations in Ukraine and Poland". *Microsoft Security Blog*, October 14, 2022. https://www.microsoft.com/en-us/security/blog/2022/10/14/new-prestige-ransomware-impacts-organizations-in-ukraine-and-poland/.

———. "Ransomware as a Service: Understanding the Cybercrime Gig Economy and How to Protect Yourself". *Microsoft Security Blog*, May 9, 2022. https://www.microsoft.com/en-us/security/blog/2022/05/09/ransomware-as-a-service-understanding-the-cybercrime-gig-economy-and-how-to-protect-yourself/.

Nakashima, Ellen, and Dalton Bennett. "A Ransomware Gang Shut down after Cybercom Hijacked Its Site and It Discovered It Had Been Hacked". *The Washington Post*, November 3, 2021. https://www.washingtonpost.com/national-security/cyber-command-revil-ransomware/2021/11/03/528e03e6–3517–11ec-9bc4–86107e7b0ab1_story.html.

———, and Dalton Bennett. "Ring of Ransomware Hackers Targeted by Authorities in United States and Europe". *The Washington Post*, November 11, 2021. https://www.washingtonpost.com/national-security/revil-ransomware-arrests-doj/2021/11/08/9432dfc2–409f-11ec-a88e-2aa4632af69b_story.html.

———, and Rachel Lerman. "FBI Held Back Ransomware Decryption Key from Businesses to Run Operation Targeting Hackers". *The Washington Post*, September 21, 2021. https://www.washingtonpost.com/national-security/ransomware-fbi-revil-decryption-key/2021/09/21/4a9417d0-f15f-11eb-a452–4da5fe48582d_story.html.

———. "Cyber Command has sought to disrupt the world's largest botnet hoping to reduce its potential impact on the election". *The Washington Post*. October 9, 2020. https://www.washingtonpost.com/national-security/cyber-command-trickbot-disrupt/2020/10/09/19587aae-0a32–11eb-a166-dc429b380d10_story.html

———. "Powerful NSA Hacking Tools Have Been Revealed Online". The *Washington Post*, August 16, 2016. https://www.washingtonpost.com/world/

national-security/powerful-nsa-hacking-tools-have-been-revealed-online/2016/08/16/bce4f974-63c7-11e6-96c0-37533479f3f5_story.html.

Narang, Satnam. "ContiLeaks: Chats Reveal Over 30 Vulnerabilities Used by Conti Ransomware—How Tenable Can Help". *Tenable Blog*, March 24, 2022. https://www.tenable.com/blog/contileaks-chats-reveal-over-30-vulnerabilities-used-by-conti-ransomware-affiliates.

National Cyber Security Centre. "Annual Review 2022: Making the UK the Safest Place to Live and Work Online". 2022. https://www.ncsc.gov.uk/files/NCSC-Annual-Review-2022.pdf.

————. "Ransomware, Extortion and the Cyber Crime Ecosystem". National Crime Agency, September 11, 2023. https://nationalcrimeagency.gov.uk/who-we-are/publications/672-ransomware-extortion-and-the-cyber-crime-ecosystem/file.

Nazarov, Denis, and Olga Emelyanova. "Blackmailer: The Story of Gpcode". *SecureList*, June 26, 2006. https://securelist.com/blackmailer-the-story-of-gpcode/36089/.

NB65. X, February 26, 2022, https://x.com/xxNB65/status/1497446425004810246

NCA: National Crime Agency. "International investigation disrupts the world's most harmful cyber crime group". February 20, 2024. https://www.nationalcrime-agency.gov.uk/news/nca-leads-international-investigation-targeting-worlds-most-harmful-ransomware-group–

————. "Russian ransomware group hit with new sanctions". September 7, 2023. https://web.archive.org/web/20240105184629/https://www.nationalcrime-agency.gov.uk/news/russian-ransomware-group-hit-with-new-sanctions–

NCSC: National Cyber Security Centre. "Encryption malware—What next?" January 5, 2022. https://www.ncsc.admin.ch/ncsc/en/home/infos-fuer/infos-unternehmen/vorfall-was-nun/ransomware.html–

Neal, Ryan W. "CryptoLocker Virus Holds Computers For Ransom". *International Business Times*, October 21, 2013. https://www.ibtimes.com/cryptolocker-virus-new-malware-holds-computers-ransom-demands-300-within-100-hours-threatens-encrypt.

Nechepurenko, Ivan. "Russia Says It Shut Down Notorious Hacker Group at U.S. Request". *The New York Times*, January 14, 2022. Accessed May 7, 2024. https://www.nytimes.com/2022/01/14/world/europe/revil-ransomware-russia-arrests.html.

Neemani, Dor, and Asaf Rubinfeld. "Diavol—A New Ransomware Used By Wizard Spider?" Fortinet, July 1, 2021. https://www.fortinet.com/blog/threat-research/diavol-new-ransomware-used-by-wizard-spider.

NewBedford. "Mayor Discusses Impact of Ransomware Attack on New Bedford's Computer System". Accessed May 5, 2024. https://www.newbedford-ma.gov/blog/news/mayor-discusses-impact-of-ransomware-attack-on-new-bedfords-computer-system/.

SELECTED BIBLIOGRAPHY

Nomios Group. "What Is REvil Ransomware?" Accessed May 5, 2024. https://www.nomios.com/resources/what-is-revil-ransomware/.

Novetta. "Operation Blockbuster: Unraveling Thread of the Sony Attack". February, 2016. https://web.archive.org/web/20160226161828/https://www.operationblockbuster.com/wp-content/uploads/2016/02/Operation-Blockbuster-Report.pdf.

O'Connell, Sean. "To Ban Ransomware Payments or Not to Ban Ransomware Payments: The Problems Drafting Legislation in Response to Ransomware". *Journal of International Business and Law, 22*(1), 2023. https://scholarlycommons.law.hofstra.edu/jibl/vol22/iss1/6–

O'Sullivan, Kevin, Georgia Edkins, and Michael Powell. "Cyber Raid Hits High Society Jeweller Graff". PressReader. *The Scottish Mail*, October 31, 2021. https://www.pressreader.com/uk/the-scottish-mail-on-sunday/20211031/281487869567973.

———. "Cyber Hackers Who Carried out Jewellers Heist Make Grovelling Apology". Mail Online, November 6, 2021. https://www.dailymail.co.uk/news/article-10172879/Russian-cyber-hackers-carried-virtual-heist-jewellers-Graff-make-grovelling-apology.html.

———. "Massive Cyber Heist Rocks High Society Jeweller Graff". Mail Online, October 30, 2021. https://www.dailymail.co.uk/news/article-10148265/Massive-cyber-heist-rocks-high-society-jeweller-Graff.html.

Office of Information Security. "HC3: Threat Profile—Threat Profile: Black Basta". *Health Sector Cybersecurity Coordination Center*, March 15, 2023. https://www.hhs.gov/sites/default/files/black-basta-threat-profile.pdf.

Pitrelli, Monica. "Leaked Documents Show Notorious Ransomware Group Has an HR Department, Performance Reviews and an 'Employee of the Month.'" CNBC, April 13, 2022. https://www.cnbc.com/2022/04/14/conti-ransomware-leak-shows-group-operates-like-normal-tech-company.html.

Price, Ned. "Reward Offers for Information to Bring Conti Ransomware Variant Co-Conspirators to Justice". United States Department of State, May 6, 2022. https://www.state.gov/reward-offers-for-information-to-bring-conti-ransomware-variant-co-conspirators-to-justice/.

PRODAFT. "CONTI Ransomware Group". Accessed May 5, 2024. https://resources.prodaft.com/conti-ransomware-group-report.

Protos Staff. "Dutch police recover 90% of victim decryption keys in ransomware scam". *Protos*. March 1, 2023. https://protos.com/dutch-police-recover-90-of-victim-decryption-keys-in-ransomware-scam/

Ransomware Task Force. "Combating Ransomware: A Comprehensive Framework for Action: Key Recommendations from the Ransomware Task Force". *IST: Institute for Security and Technology*. 2021. https://securityandtechnology.org/wp-content/uploads/2021/09/IST-Ransomware-Task-Force-Report.pdf.

———. "Ransomware Task Force: Doubling Down". *IST: Institute for Security and Policy*. 2024. https://securityandtechnology.org/virtual-library/reports/ransomware-task-force-doubling-down/.

SELECTED BIBLIOGRAPHY

Rapid7 Blog. "Conti Ransomware Group Internal Chats Leaked". March 1, 2022. https://www.rapid7.com/blog/post/2022/03/01/conti-ransomware-group-internal-chats-leaked-over-russia-ukraine-conflict/.

Raywood, Dan. "Proposals to Ban Ransomware Payments Rumoured". *SC Media UK*. May 22, 2024. https://insight.scmagazineuk.com/proposals-to-ban-ransomware-payments-rumoured.

Reddick, James. "'Prolific' crypto money laundering platform ChipMixer shuttered by Germany, US". *The Record*. March 15, 2023. https://therecord.media/chipmixer-takedown-cryptocurrency-money-laundering-europol-doj.

Reed, Jonathan. "How Reveton Ransomware-as-a-Service Changed Cybersecurity". *Security Intelligence*, December 19, 2022. https://securityintelligence.com/articles/how-reveton-raas-changed-cybersecurity/.

Rob Joyce, *"Disrupting State Hackers"*. USENIX Enigma, 2016, https://www.usenix.org/sites/default/files/conference/protectedfiles/engima2016_transcript_joyce_v2.pdf.

Roccia, Thomas. "Using Python to Unearth a Goldmine of Threat Intelligence from Leaked Chat Logs". *Jupyter Security Break*. Accessed May 5, 2024. https://jupyter.securitybreak.io/Conti_Leaks_Analysis/Conti_Leaks_Notebook_TR.html.

Roncone, Gabby, Dan Black, John Wolfram, Tyler McLellan, Nick Simonian, Ryan Hall, Anton Prokopenkov, Dan Perez, Lexie Aytes, and Alden Wahlstrom. *APT44: Unearthing Sandworm*. Mandiant, 2024.

Roth, Florian, Pasquale Stirparo, David Bizeul, Brian Bell, Ziv Chang, Joel Esler, Kristopher Bleich, et al. "APT Groups and Operations". March 2020. https://docs.google.com/spreadsheets/d/1H9_xaxQHpWaa4O_Son4Gx0YOIzlcBWMsdvePFX68EKU/edit#gid=1864660085.

Russo, Kristopher. "Threat Assessment: Luna Moth Callback Phishing Campaign". *Unit 42*, November 21, 2022. https://unit42.paloaltonetworks.com/luna-moth-callback-phishing/.

SafeGuard Cyber Team. "BazarCall: A Double-Edged Attack". Safeguard Cyber, September 7, 2022. https://www.safeguardcyber.com/blog/security/3-bazar-call-phishing-groups-wreaking-havoc.

Sanger, David E. "Russia's Most Aggressive Ransomware Group Disappeared. It's Unclear Who Made That Happen". *The New York Times*, July 13, 2021. https://www.nytimes.com/2021/07/13/us/politics/russia-hacking-ransomware-revil.html.

Santanna, Jair. "The COMPLETE Translation of Leaked Files Related to Conti Ransomware Group". *GitHub*. Accessed May 5, 2024. https://github.com/NorthwaveSecurity/complete_translation_leaked_chats_conti_ransomware.

Schaffer, Aaron. "Ransomware Hackers Have a New Worst Enemy: Themselves". *The Washington Post*, October 12, 2022. https://www.washingtonpost.com/politics/2022/10/12/ransomware-hackers-have-new-worst-enemy-themselves/.

Schwartz, Mathew J. "REvil's Ransomware Success Formula: Constant Innovation".

SELECTED BIBLIOGRAPHY

Bank Info Security, July 2, 2021. https://www.bankinfosecurity.com/revils-ransomware-success-formula-constant-innovation-a-16976.

Secureworks. "Arhiveus Ransomware Trojan Threat Analysis". Accessed May 4, 2024. https://www.secureworks.com/research/arhiveus. September 10, 2023. https://x.com/HaciendaCR/status/1516190939114803203.

Shapiro, Jacob N. *The Terrorist's Dilemma: Managing Violent Covert Organizations*. Princeton University Press, 2013. http://www.jstor.org/stable/10.2307/j.ctt2tt8v9.

Shier, John, Mat Gangwer, Greg Iddon, and Peter Mackenzie. "The Active Adversary Playbook 2021". *Sophos News*, May 18, 2021. https://news.sophos.com/en-us/2021/05/18/the-active-adversary-playbook-2021/.

Shriebman, Yaara. "Ransomware 2021—The Bad, The Bad & The Ugly". Cyberint, December 30, 2021. https://cyberint.com/blog/research/ransomware-2021-the-bad-the-bad-the-ugly/.

———. "To Be CONTInued? Conti Ransomware Heavy Leaks". Cyberint, March 9, 2022. https://cyberint.com/blog/research/contileaks/.

Singh, Simon. *The Code Book: The Science of Secrecy from Ancient Egypt to Quantum Cryptography*. Random House, 1999.

Smeets, Max. "Building a Cyber Force Is Even Harder Than You Thought". War on the Rocks, May 12, 2022. https://warontherocks.com/2022/05/building-a-cyber-force-is-even-harder-than-you-thought/.

———. "The Role of Military Cyber Exercises: A Case Study of Locked Shields". edited by T. Jančárková, G. Visky, and I. Winther, 2022.

———. *No Shortcuts: Why States Struggle to Develop A Military Cyber-Force*. Oxford University Press, 2022.

Smilyanets, Dmitry. "'I Scrounged through the Trash Heaps ... Now I'm a Millionaire:' An Interview with REvil's Unknown". *The Record*, March 16, 2021. https://therecord.media/i-scrounged-through-the-trash-heaps-now-im-a-millionaire-an-interview-with-revils-unknown.

Soni, Anuj, and Ryan Chapman. "The Curious Case of 'Monti' Ransomware: A Real-World Doppelganger". *BlackBerry Blog*, September 7, 2022. https://blogs.blackberry.com/en/2022/09/the-curious-case-of-monti-ransomware-a-real-world-doppelganger.

Sophos. "Maze Ransomware: Extorting Victims for 1 Year and Counting". *Sophos News*, May 12, 2020. https://news.sophos.com/en-us/2020/05/12/maze-ransomware-1-year-counting/.

Stolyarov, Vlad, and Benoit Sevens. "Exposing Initial Access Broker with Ties to Conti". Google, March 17, 2022. https://blog.google/threat-analysis-group/exposing-initial-access-broker-ties-conti/.

Sunny. "NB65: Riesen-Blamage mit falschem Kaspersky Source Code-Leak". Tarnkappe.info, March 10, 2022. https://tarnkappe.info/artikel/nb65-riesen-blamage-mit-falschem-kaspersky-source-code-leak-221205.html.

Syal, Rajeev. "Ransomware Attacks in UK Have Doubled in a Year, Says GCHQ Boss".

SELECTED BIBLIOGRAPHY

The Guardian, October 25, 2021. https://www.theguardian.com/uk-news/2021/oct/25/ransomware-attacks-in-uk-have-doubled-in-a-year-says-gchq-boss.

TechNewsWorld. "Sophos Cracks Archiveus Ransomware Code". June 2, 2006. https://www.technewsworld.com/story/sophos-cracks-archiveus-ransomware-code-50881.html.

Temple-Raston, Dina. "Ransomware Diaries: Undercover with the Leader of LockBit". The Record, January 16, 2023. https://therecord.media/ransomware-diaries-undercover-with-the-leader-of-lockbit.

The DFIR Report. "Ryuk's Return". October 8, 2020. https://thedfirreport.com/2020/10/08/ryuks-return/.

The Parmak. "Conti-Leaks-Englished". GitHub. Accessed May 5, 2024. https://github.com/TheParmak/conti-leaks-englished.

The Ransomware Files Podcast. *The Ransomware Files Podcast, Episode 6: Kaseya and REvil*. Youtube. April 8, 2022. https://www.youtube.com/watch?v=dO8hNhi9WmM.

The White House. "Joint Statement of the Ministers and Representatives from the Counter Ransomware Initiative Meeting October 2021". October 14, 2021. https://www.whitehouse.gov/briefing-room/statements-releases/2021/10/14/joint-statement-of-the-ministers-and-representatives-from-the-counter-ransomware-initiative-meeting-october-2021/.

Thomas, William. "WizardSpider Using Legitimate Services as Cloak of Invisibility". *CYJAX*, April 20, 2021. https://www.cyjax.com/wizardspider-using-legitimate-services-as-cloak-of-invisibility/.

Threat Hunter Team. "The Evolution of Emotet: From Banking Trojan to Threat Distributor". *Symantec Enterprise Blogs: Threat Intelligence*, July 18, 2018. http://prod-blogs-ui.client-b1.bkjdigital.com/blogs/threat-intelligence/evolution-emotet-trojan-distributor.

Trend Micro DE. "New Stampado Ransomware Sold Cheap on the Dark Web". July 14, 2016. https://www.trendmicro.com/vinfo/de/security/news/cybercrime-and-digital-threats/new-stampado-ransomware-sold-cheap-on-the-dark-web.

Trend Micro Research. "Ransomware Spotlight: Black Basta". Trend Micro, September 1, 2022. https://www.trendmicro.com/vinfo/us/security/news/ransomware-spotlight/ransomware-spotlight-blackbasta.

————. "Ransomware Spotlight: BlackByte". Trend, July 5, 2022. https://www.trendmicro.com/vinfo/us/security/news/ransomware-spotlight/ransomware-spotlight-blackbyte.

TRM Insights. "TRM Analysis Corroborates Suspected Ties Between Conti and Ryuk Ransomware Groups and Wizard Spider". April 6, 2022. https://www.trmlabs.com/post/analysis-corroborates-suspected-ties-between-conti-and-ryuk-ransomware-groups-and-wizard-spider.

Tsialemis, Kostas. "TRANSLATED Conti Leaked Comms". GitHub, 2022. https://github.com/tsale/translated_conti_leaked_comms.

Uberti, David. "Russia-Linked Ransomware Groups Are Changing Tactics to Dodge

SELECTED BIBLIOGRAPHY

Crackdowns". *The Wall Street Journal*, June 2, 2022. https://www.wsj.com/articles/russia-linked-ransomware-groups-are-changing-tactics-to-dodge-crackdowns-11654178400.

Umawing, Jovi. "Karakurt Extortion Group: Threat Profile". *Threat Intelligence.* ThreatDown by Malwarebytes, June 14, 2022. https://www.threatdown.com/blog/karakurt-extortion-group-threat-profile/.

United States Department of Justice. "Emotet Botnet Disrupted in International Cyber Operation". January 28, 2021. https://www.justice.gov/opa/pr/emotet-botnet-disrupted-international-cyber-operation.

————. "Justice Department Investigation Leads to Takedown of Darknet Cryptocurrency Mixer that Processed Over $3 Billion of Unlawful Transactions". March 15, 2023. https://www.justice.gov/usao-edpa/pr/justice-department-investigation-leads-takedown-darknet-cryptocurrency-mixer-processed.

————: Office of Public Affairs. "Department of Justice Seizes $2.3 Million in Cryptocurrency Paid to the Ransomware Extortionists Darkside". June 7, 2021. https://www.justice.gov/opa/pr/department-justice-seizes-23-million-cryptocurrency-paid-ransomware-extortionists-darkside.

————: Office of Public Affairs. "Latvian National Charged for Alleged Role in Transnational Cybercrime Organization". June 4, 2021. https://www.justice.gov/opa/pr/latvian-national-charged-alleged-role-transnational-cybercrime-organization.

————: Office of Public Affairs. "U.S. Leads Multi-National Action Against 'Gameover Zeus' Botnet and 'Cryptolocker' Ransomware, Charges Botnet Administrator". June 2, 2014. https://www.justice.gov/opa/pr/us-leads-multi-national-action-against-gameover-zeus-botnet-and-cryptolocker-ransomware.

United States Department of the Treasury. "Treasury Sanctions Russia-Based Hydra, World's Largest Darknet Market, and Ransomware-Enabling Virtual Currency Exchange Garantex". April 5, 2022. https://home.treasury.gov/news/press-releases/jy0701.

————. "U.S. Treasury Issues First-Ever Sanctions on a Virtual Currency Mixer, Targets DPRK Cyber Threats". May 6, 2022. https://home.treasury.gov/news/press-releases/jy0768.

————. "United States and United Kingdom Sanction Members of Russia-Based Trickbot Cybercrime Gang". February 9, 2023. https://home.treasury.gov/news/press-releases/jy1256.

Vaas, Lisa. "Conti Ransomware Decryptor, TrickBot Source Code Leaked—Vulnerability Database". Vulners Database, March 2, 2022. https://vulners.com/threatpost/THREATPOST:0B290DDF3FE14178760FDC2229CB1383.

————. "Conti Ransomware V. 3, Including Decryptor, Leaked". Threatpost, March 21, 2022. https://threatpost.com/conti-ransomware-v-3-including-decryptor-leaked/179006/.

Vallace, Chris, and Joe Tidy. "Ransomware Attack hits Dozens of Romanian

Hospitals". *BBC News*. February 13, 2024. https://www.bbc.com/news/technology-68288150.

Vedere Labs. "Analysis of Conti Leaks". *Forescout*, March 2022. https://www.forescout.com/resources/analysis-of-conti-leaks/.

Volz, Dustin, and Sarah Young. "White House Blames Russia for 'Reckless' NotPetya Cyber Attack". *Reuters*, February 16, 2018. https://www.reuters.com/article/idUSKCN1FZ0PR/.

Vx-underground. X, July 12, 2021. https://x.com/vxunderground/status/1414588622670532616.

Waldman, Arielle. "Distrust, Feuds Building among Ransomware Groups". TechTarget: Security, February 3, 2022. https://www.techtarget.com/searchsecurity/news/252512902/Distrust-feuds-building-among-ransomware-groups.

Walker, Amy. "UK '95% Sure' Russian Hackers Tried to Steal Coronavirus Vaccine Research". *The Guardian*, July 17, 2020. https://www.theguardian.com/world/2020/jul/17/russian-hackers-steal-coronavirus-vaccine-uk-minister-cyber-attack.

Walter, Jim. "From the Front Lines—3 New and Emerging Ransomware Threats Striking Businesses in 2022". *SentinelOne*, June 22, 2022. https://www.sentinelone.com/blog/from-the-front-lines-3-new-and-emerging-ransomware-threats-striking-businesses-in-2022/.

Waqas. "Anonymous NB65 Claims Hack on Russian Payment Processor Qiwi". *HackRead*, May 9, 2022. https://www.hackread.com/anonymous-nb65-hacki-russia-payment-processor-qiwi/.

Wendel, Anton. "Emotet Harvests Microsoft Outlook". *Cyber.WTF*, October 12, 2017. https://cyber.wtf/2017/10/12/emotet-beutet-outlook-aus/.

Weston, Sabina. "Irish Police Seize Conti Domains Used in HSE Ransomware Attack". ITPro, September 6, 2021. https://www.itpro.com/security/ransomware/360786/irish-police-seize-conti-domains-used-in-hse-ransomware-attack.

Weston, Sabrina. "Irish police seize Conti domains used in HSE ransomware attack". *ITPro*. September 6, 2021. https://www.itpro.com/security/ransomware/360786/irish-police-seize-conti-domains-used-in-hse-ransomware-attack.

White, Geoff. *The Lazarus Heist: From Hollywood to High Finance: Inside North Korea's Global Cyber War*. Penguin Business, 2022.

Whitworth, Damian. "How I Exposed Alexei Navalny's Poisoners and Fell Foul of Putin". *The Times*, March 4, 2023. sec. News. https://www.thetimes.co.uk/article/bellingcat-eliot-higgins-alexei-navalny-poisoning-putin-russia-ukraine-jvsmpqmdb.

Wikileaks. "Vault 7: CIA Hacking Tools Revealed". March 7, 2017. https://wikileaks.org/ciav7p1/.

Wilding, Edward. "The Authoritative International Publication on Computer Virus Prevention, Recognition and Removal". *Virus Bulletin*, March 1990. https://www.virusbulletin.com/uploads/pdf/magazine/1990/199003.pdf.

———. "The Authoritative International Publication on Computer Virus

SELECTED BIBLIOGRAPHY

Prevention, Recognition and Removal". *Virus Bulletin*, January 1992. https://www.virusbulletin.com/uploads/pdf/magazine/1992/199201.pdf.

Wolf, Arctic. "The Karakurt Web: Threat Intel and Blockchain Analysis". *Arctic Wolf*, April 15, 2022. https://arcticwolf.com/resources/blog/karakurt-web/.

INDEX

Note: Page numbers followed by "*n*" refer to notes, "*t*" refer to tables, "*f*" refer to figures.

INDEX

INDEX

INDEX